A Nomadic Pedagogy about Technology

# International Technology Education Studies



VOLUME 18

The titles published in this series are listed at *brill.com/ites*

# A Nomadic Pedagogy about Technology

*Teaching the Ongoing Process of Becoming Ethnotechnologically Literate*

*By*

John R. Dakers

BRILL

LEIDEN | BOSTON

Cover illustration: iStock.com/Toa55

All chapters in this book have undergone peer review.

The Library of Congress Cataloging-in-Publication Data is available online at https://catalog.loc.gov

Typeface for the Latin, Greek, and Cyrillic scripts: "Brill". See and download: brill.com/brill-typeface.

ISSN 1879-8748
ISBN 978-90-04-53698-2 (paperback)
ISBN 978-90-04-53699-9 (hardback)
ISBN 978-90-04-53700-2 (e-book)



This book is printed on acid-free paper and produced in a sustainable manner.

*For Anna and Robin*
*And all young people everywhere*
*For it is they, not we, who must reimagine their future*

# Contents

# Preface

History tells us that there has always been a relationship between human beings and their technologies. Anthropological findings offer evidence from pre-history, whilst history, which according to (Kramer, 1956), begins at Sumer, suggests that it was the Sumerians who gave us the first cities and perhaps most crucially, invented writing in the form of cuneiform. This, for the first time, gave rise to written accounts of life at that time. Whatever their origins, however, it is evident that from the time human beings occupied this Earth, they have managed to evolve, from that day to this, only as a result of their relationship with technologies. An understanding of the concept of technology is, therefore, essential if we are to comprehend the world both of today and of the past and into the future. We have, moreover, now entered the era known as the Anthropocene, an era that, as a result of our over-exploitation of the natural resources of the planet for economic technological gain, has led to what is generally considered to be the greatest existential threat to human existence ever encountered. As governments try to balance economic growth through consumerism, with a scaling down of the use of fossil fuels, for example, the damage continues as a cancerous growth which is slowly, but inexorably, destroying the entire eco-system. As I write this, there is a war being fought in Ukraine, one that has been initiated by Russia. This war is having a devastating effect on energy supplies throughout the world where costs are escalating at hitherto unprecedented levels. The results of this war serve as a stark reminder of the world's heavy reliance on fossil fuels and reveals the urgent need to rapidly change the way our energy is created. It is vital, therefore, that members of the next generation be given a voice through which they cannot only express, their concerns, but which will enable them to become active participants in the discovery of possible solutions to this crisis. It is, after all, they who will inherit the world.

At present technology education in schools does not offer a platform where young people can express themselves in this context. Vocationally orientated technology education continues, for the most part, to offer craft-based skill development and related knowledges that are in the service of industry. This, despite the fact that no one can know in advance, what these industry requirements might be. Whilst curricula for technology education does support the inclusion of technological literacy, this is, for the most part, located within the extant pedagogies applied in technology education. Technological literacy, therefore, is in reality, rarely, if ever, covered. What is needed is the creation of a modern-day Agora where young people can express their views on the technologically textured world they inhabit and will inherit. This will require the development of a new form of pedagogy: a nomadic pedagogy that will enable young people and teachers alike to consider the ways in which technology affects them in everyday life, and to express the ways in which they think they themselves affect technological development. Rather than trying to resurrect technological literacy, which has been tried again and again over the years to no lasting avail, I will propose that a new subject entitled ethnotechnological literacy needs to be developed, utilising a new nomadic pedagogy.

It was around the late 1980s that the introduction of the concept of technological literacy began to be seriously considered within the group of subjects variously known as technology education, design and technology education and technical education. This innovation was designed to incorporate new, social and philosophical dimensions into a subject that was considered to be too craft orientated and vocational in nature. Technological literacy was, as de Vries articulates, designed to develop in young people, an ability not only to critique technology, but also to enable them "to live and work in a technological society by making responsible and sophisticated use of technology" (de Vries, 2018a, p. 78). This book will explore the development of technological literacy since its inception up to the present day and conclude by advancing a new nomadic pedagogy for teaching about technology through the ongoing process of becoming ethnotechnologically literate. I use the prefix 'ethno,' which means people, society and culture, as a means of differentiating between the concept of technology considered in purely objective terms, towards a consideration of technology in sociocultural terms. I make this distinction because 'technological literacy' is often perceived in objective terms, whereas the addition of 'ethno' clarifies ethnotechnological literacy as a subjective, ergo sociocultural concept. It is thus the study of the human technology relationship.

I want also to stress at this point that the concerns that I will raise in this book are related to technology education programmes that are vocationally orientated and which have a strong focus on craft skills development. I believe these programmes to be predominant. I acknowledge that there are examples of technology education delivery that go beyond this purely craft-based, vocational model. These more progressive programmes usually incorporate design, and some even attempt to explore ethical dimensions of the technologically textured world we inhabit. Even within these, however, the concept of ethnotechnological literacy does not feature to the extent I believe it should, especially when we consider the technologically orientated, world-wide problems we face today. My experience as a teacher, university lecturer and researcher of technology education, is reinforced by ongoing expressions of concern, in both written and anecdotal form, as expressed by colleagues from around the world.

This leads me to argue that the dominant orthodoxy, in the public perception and in the delivery of technology education, remains stubbornly within the vocationally orientated mould, with an emphasis on craft-based occupations. It is for that reason that I use the term 'craft-based technology education' as distinct from 'education *about* technology' which is more in alignment with the concept of ethnotechnological literacy. Philosophy explores and informs the way we understand concepts such as pedagogy and curriculum, as well as our understanding of the relationship between technology and humans, and between technology and the environment.

It is important to understand that education *about* technology also operates in a political environment, one that informs policies that in turn impact upon pedagogy and curriculum development. Thus, the concept of ethnotechnological literacy serves as an ideal platform from which to consider these important issues, especially in today's technologically textured world. I have chosen to explore the philosophy of Gilles Deleuze who, in association with Felix Guattari, introduced the concept of 'nomadology' to the philosophical field. Relatively recently, other philosophers, influenced by the work of Deleuze, have begun to explore the concept of nomadic education. Influenced by this poststructuralist philosophy, I will offer a completely new and unique perspective on education about technology, through the development of a nomadic pedagogy related to the ongoing process of becoming ethnotechnologically literate, as outlined in the title of this book.

## Acknowledgements

I well remember attending my first ITEA conference in Columbus, Ohio, in March 2002. This was where I had my first encounter with Marc de Vries, one of the founders of the international PATT conferences. I also remember having a conversation with him in the hotel bar one evening, when I asked him why the PATT conferences had never been held in Glasgow, my home city. "Good point," he said, "Why don't you organise it for next year?" I remember saying "OK, I will." I remember then thinking to myself, "Never again have a conversation with Marc de Vries whilst in a bar!" That was 20 years ago and it was one of the most important meetings I have ever had. It quite literally changed my academic life. The 2003 PATT conference was an amazing success and I have, ever since, developed a very strong relationship with Marc who has not only mentored me through my research, supported me through some difficulties with my university, co-edited several books and book chapters with me (and other colleagues), but has also become a dear friend. Marc kindly offered to read my initial transcript of this book and made some helpful observations which I have incorporated. I cannot express enough my sincere thanks for the incredible support and advice he gave to me in the production of this, my first monograph.

I also wish to thank Wendy, my lovely lady, whose husband I have had the privilege of being for many years. Wendy has not only supported my research but carried out and published her own research projects related to technology education as well as many joint projects with me. We worked together for several years on two longitudinal research studies with the European Commission on science and technology education. This gave us the opportunity to visit many European countries and study their science and technology education projects. This book could not have been accomplished without her support and what I refer to as a sprinkling of her fairy dust as she proofreads and sorts out my punctuation, grammar, and very long sentences. Thank you so much Wendy.

I could not have managed to put this book together without the incredible support of Brill. I would especially like to thank John Bennett, Katerina Sofianou and Jolanda Karada who have all been so supportive in this endeavour. Many thanks to you all.

Silly as it may sound, I would also like to thank all the philosophers I have read who have helped bring me to this place. I especially want to give posthumous thanks to Gilles Deleuze, Félix Guattari, Gilbert Simondon and Bernard Stiegler for their wonderful, albeit difficult, philosophical insights, to which I hope I have done justice.

# Figures

CHAPTER 1

# Introduction

## *Setting the Scene*

> Perdu dans ce vilain monde, coudoye par les foules, je suis comme un homme lasse dont Poeil ne voit en arriere, dans les annees profondes, que desabusement et amertume, et devant lui qu'un orage ou rien de neuf n'est contenu, ni enseignement, ni douleur.
> [Lost in a wasteland, jostled by the crowds, I am like a weary man who sees in the depths of the past behind him, nothing but disappointment and bitterness, and before him a storm that contains nothing new, neither insight, nor grief.]
>
> CHARLES BAUDELAIRE (1846, p. 1744)

∵

I am in agreement with Holland, when he empathises with Baudelaire, by stating "c'est moi! For I, too, feel like someone who sees little but bitter disappointment in the past, like someone being blown irresistibly backwards into the future" (Holland, 2006, p. XI). I too am someone who can only look aghast at how little technology education today reflects the issues relating to the technologically textured lifeworld we currently inhabit. O'Riley, in her book *Technology, Culture, and Socioeconomics*, also offers a somewhat dystopic but nevertheless accurate account of the frustration, in this case by a female student undergoing her initial teacher education course in technology education. The student asks:

> How has it come to be that in spite of recent revisions, technology education remains limited to technical and trades-orientated technologies? How has it come to be that a critical and urgent conversation on gender, cultural, socioeconomic, global, and environmental issues in relation to technology is not at the foreground of technology curricula? How has it come to be that technologies associated with every-day living and the home have been pushed to the side? How has it come to be that western technologies are languaged as high, and indigenous technologies talked about as low, or no, technologies? How is it that there is such an uncritical

 | DOI:10.1163/9789004537002_001

> emphasis on design and making when the earth is suffocating, gasping for breath, amidst the unparalleled production, consumption, and waste of the past fifty years? What other conversations on technologies need to become a part of the technological discourses in education today? Who might be able to bring these conversations to the table? (O'Riley, 2003, p. 3)

It is the policy makers who decide at a national level what technology education should be, and this is usually linked to the perceived needs of industry. Then we have the curriculum designers who prescribe the learning standards, objectives and assessment protocols on behalf of governments. Academic researchers offer insights that seek to herald change into what is now considered by them to be a subject that is out of date, although alas, these insights are mostly ignored. Universities that offer initial teacher education courses on the subject, lack the appropriate academic rigour.[1] As a consequence, student teachers spend a great deal of time learning the skills, mostly craft-based, that they will then be expected to pass on, in turn, to their students at school. Universities do not consider qualifications in technology subjects to be eligible for entry into engineering or architectural degree courses. The consequence of this is to force the subject, as an openly vocationally oriented one, towards those interested in craft-based occupations. In addition, head teachers make decisions about resource allocation. This can be expensive for technology education, which as a result, is likely to be pared back over time. Technology teachers who deliver the subject, many of whom are reluctant to herald change, prefer instead to repeat the same old workshop lessons they have been conducting for years. Finally, the students of technology education are never expected to read any books about the history, philosophy, or impact that technology has on societies, because 'tech ed' is perceived to be the subject chosen predominantly by boys, especially those less academically inclined. The subject is, and has always been, considered to be an initial training ground designed to serve the perceived needs of industry (see, for example Toft, 2007, and Weale, 2021). These diverse perspectives all conspire to work against each other, the inevitable outcome being that nothing ever changes (see also Dakers, 2018; Barlex, 2018; Milne, 2018; Petrina, 2018; Williams, 2018).

In the preface to his book *Advanced Teaching Methods for the Technology Classroom*, Stephen Petrina outlines three areas he considers as being crucial to the teaching of technology:

> when we teach about technology, we are dealing with content and dispositions; when we teach through technology, we are dealing with processes

> and skills; teaching for technology refers to occupations and roles. By themselves, none of these three orientations provides an adequate education. A conscious integration of the three orientations is what differentiates the approach in this book from more narrow studies about technology or training for technology. Effective technology teaching requires that we balance these three orientations. (Petrina, 2007, pp. IX–X)

Petrina puts forward a perfectly reasonable case by arguing, essentially, that technology education is in much need of significant reform. It needs to be much more than learning about the processes and skills that serve the perceived needs of industry. However, these arguments, as presented by Petrina, have continued for decades, to no avail. Integrating teaching *about* technology, which for me is synonymous with 'technological literacy,' into the technology education curriculum, continues to be stubbornly unattainable. This is despite a wealth of peer-reviewed academic research over the years that has been advocating exactly what Petrina is arguing for, above (see, for example, *New Frontiers in Technology Education* by Dakers, 2014, *The Future of Technology Education* by Williams et al., 2015, and the *Handbook of Technology Education* by de Vries, 2018a, 2018b). Could it be that the type of pedagogy employed in the presentation of technology education is diametrically opposed to the type of pedagogy that is crucial for teaching *about* technology? For many years now, research into pedagogy generally, and technology education specifically, has been calling for change. However, change comes slowly in educational environments and change in the pedagogy employed in technology education is no different. It is for that reason that I will present a new and radical approach for an education *about* technology, as well as a corresponding new nomadic pedagogy in this book. Whenever I use the term education *about* technology, which I will later propose to be made manifest as an ongoing process of becoming ethnotechnologically literate, I mean technology education as a transformative process of change, one of becoming other. This is in contrast to the extant model of craft-based technology education; one that is reduced to the unidirectional transmission from teacher to student of a pre-given curriculum. This is, indeed, more akin to training (Semetsky, 2017, p. 11). I will argue that whilst this pedagogical framework continues to be the dominant orthodoxy in craft-based technology education workshops or laboratory settings, it is not at all suited to the ongoing process of becoming ethnotechnologically literate, a process that requires an altogether different form of pedagogy. The educational paradigm that I propose aspires to be a transformative and liberating experience that does not stop at the school gates. It is education about technology that involves self-development and self-cultivation. It involves learning of the type the German term 'Bildung'[2] expresses best. As Stoltz (2021, p. 199, note 5) explains.

> [I]t has multiple meanings – such as 'education,' 'culture,' 'formation,' 'fashioning,' 'shaping,' 'setting up' – none of which captures its significance, and so translation into the English word of 'education' would be an injustice to the conceptual richness of Bildung because it is not just education, but involves much more, such as self-cultivation, self-development, self-formation and so on.

I believe that in any educational setting the concept of Bildung should be considered the basis for any pedagogy employed. I particularly see it in relation to education *about* technology. In order to accommodate this, the development of ethnotechnological literacy must involve a complete change from the current pedagogy of transmission, to one that is more critically orientated, one that involves:

> a recognition that education [about technology] was important not only for gainful employment but also for creating the formative culture of beliefs, practices, and social relations that enable individuals to wield power, learn how to govern, and nurture a democratic society that takes equality, justice, shared values, and freedom seriously ... pedagogy is central to politics in that it is involved in the construction of critical agents and provides the formative culture that is indispensable to a democratic society. (Giroux, 2020, p. 2)

For this to be achieved, a nomadic pedagogy for ethnotechnological literacy must recognise the rapid changes that are taking place in today's technologically textured world. A people yet to (be)come, needs to be literate, and for the purpose of this book, be ethnotechnologically literate. Extant research into how this might be achieved, some of which will be revealed throughout this book, suggests, however, that education about technological literacy, in its present incarnation, is either completely missing, or is so seriously out of step with the current rapid and immense tectonic technological transformations, that it does not come anywhere close to fulfilling this need. To enter into a process of becoming ethnotechnologically literate, schools, both public and private, religious and non-religious, primary and secondary, will have to incorporate pedagogies that are predicated upon the concepts of Bildung, transversality and nomadicity, processes that I will go on to elucidate further, as the book progresses.

However, before that can be accomplished it is important first to define, in clear terms, what any institution, indeed any society, considers the purpose of education to be. I have lost count of the number of times I have listened to politicians wax lyrically about why a good education is so important. They do so, however, without ever defining what they believe constitutes a 'good' or 'bad' education, other than to couch it in terms of examination passes. The present curriculum content of technology education in this respect is paradoxical. The documentation claims, on the one hand, that technological literacy is an important curricular element, whilst on the other hand, excludes any serious attempt to assess it. In this assessment by examination culture, teachers, who are also assessed on the percentage of their students who successfully achieve good grades, understandably, avoid teaching technological literacy, preferring instead to devote time to the assessable components. The development of technological literacy is, in addition, theoretical and ideological in nature and so difficult to teach because it is, for the most part, not easily, if not impossible to assess summatively. It is therefore given little prominence in a craft-based orientated subject.[3]

While technology evolves at an exponential pace, technology education remains grounded in the past. The issues raised in this opening chapter will form the foundational architecture from which this book will build. In so doing, I will present my concept of what constitutes a progressive and radical critical pedagogy for the delivery of education *about* technology through the lens of the ongoing process of becoming ethnotechnologically literate. This will involve exploring what the characteristics of a nomadic pedagogy for teaching about technology involves.

## 1 A Brief Outline of the Chapters

An introduction is given in Chapter 1.

In Chapter 2 I will analyse the differences that exist between the concept of technology and that of technique. This is an important distinction that can and does cause some confusion, particularly for education *about* technology. It becomes evident that from as far back as the ancient Greeks, the debates surrounding what constituted techne then, have not particularly evolved over time. Indeed, I have in this chapter made a distinction between 'craft-based technology education' as the acquisition of craft skills and techniques, and 'education *about* technology' as the development of ethnotechnological literacy; that is, learning to be more critical and philosophical about technology by exploring the socio-cultural aspects. In order to pursue either, it is crucial that both teachers and students are able to articulate what exactly they mean when they talk about technology and technique. In so doing, the distinction between these two perspectives becomes more conspicuous, resulting in a better awareness of what constitutes technological knowledge as outlined by Carl Mitcham in his book *Thinking through Technology*. This then clarifies what pedagogy is most suitable for delivering the subject. I then cite Mitcham's theory of volition as the formation of technological knowledge that best represents my concept of the ongoing process of becoming ethnotechnologically literate. I then offer my definition of ethnotechnological literacy to complete the chapter.

In Chapter 3 I will argue that technology education in its present form cannot articulate with education *about* technology. They are essentially two different subjects involving two completely different styles of pedagogy. Technology education relies upon the development of declarative and procedural knowledges. Assessment is straightforward and based upon the precision shown in the application of each student's technique, in addition to the extent of their knowledge about tools and methods of fabrication. It is primarily quantitative in nature and teacher directed. Education *about* technology, on the other hand, lacks most of this quantitative dimension. It is qualitative in its arrangement, has no prescribed answers to any questions or issues raised, and can only truly be delivered democratically with the teacher acting as facilitator.

I start Chapter 4 by discussing the principal philosophers who have influenced or been influenced by Deleuze. I also offer an explanation as to why I have been influenced by the philosophy of Deleuze to inspire this book. One of the principle philosophical concepts articulated by Deleuze is that of the ongoing process of the 'becoming of being.' This forms one the cornerstones of this book. Human beings have for a long time considered identity only through a lens that depicts a stable and enduring state. Both Deleuze and Simondon challenge this perspective and offer instead a philosophy of the becoming of being and pre-individuation respectively. I show how these corresponding philosophies serve to inform and illustrate the argument that the relationship between human beings and their technologies is a dynamic, ever-changing process, quite the opposite of stable and enduring.

In Chapter 5 I look at the way philosophers have, over time, considered the relationship between humans and their technologies, a central theme of becoming ethnotechnologically literate. What becomes evident is that these perspectives have clearly led to both technological and socio-cultural changes in the human technology relationship, one that drives academics like Moravec, to make the claim that it will soon be possible to download human consciousness into a computer. Considerations such as these can, and do, inspire engineers to pursue these possibilities.

It is in this respect that I question what it is that constitutes being human and ask how much technology has directed the evolution of human being since they first came to inhabit the world, and vice versa in terms of technology. The conclusion that I initially draw is that neither one can be considered without the other. In other words, to be human is to be technological, and to be technological is to be human. However, to be either does not constitute a stable and enduring identity. To that end I reformulate my conclusion to argue that both are dynamic states that are contingent upon each other. The chapter raises questions that have been debated over time immemorial and highlights, as Ihde postulates, the essence of technology as being a "certain way of experiencing, relating to and organising the way humans relate to the natural world" (Ihde, 1983, p. 29). This highlights the kind of issues that have been debated over time as well as the kind of debates that a nomadic education can bring about in the ongoing process of becoming ethnotechnologically literate.

In Chapter 6 I introduce, in more detail, the philosophical concepts of nomadism, transversality, the virtual, the fold and subjectivity as crucial aspects in teaching and learning the ongoing process of becoming ethnotechnologically literate. I highlight these philosophical concepts, proffered by Deleuze, as being necessary and supplementary to the previous chapters, and so serve as a philosophical bridge to the next chapter on nomadic pedagogy.

In Chapter 7 it transpires that over thirty years ago five characteristics for technology education were proposed by Marc de Vries. I have adopted all but one of these characteristics, changed one slightly and added a new one. However, whilst these characteristics have occupied the research field for a long time, they have never become actualised in the classroom in terms of technological literacy. It is my belief that the reason for this is because the pedagogy employed in the delivery of craft-based technology education mitigates against this. In order to expedite these characteristics, therefore, a new and progressive method of teaching *about* technology must be developed in the form of a nomadic pedagogy made actual as part of an ongoing process of becoming ethnotechnologically literate. This chapter will seek to outline, in detail, what these characteristics are.

Chapter 8 presents some concluding remarks and a final argument for a progressive education.

## Notes

1 See O'Riley (2003). I also make this claim based upon my own experiences as both a mature student teacher, a teacher and then a university lecturer in Educational Studies. See also, for example: (Sykes, Bird, & Kennedy, 2010; Rani, 2017; Jamwal, 2012).

2 "The nuances of *Bildung* are notoriously difficult to replicate in English. The verb *Bilden* (from *Bild*: a picture) means to form (as after an image), and *Bildung* tends much more strongly than its English counterparts to imply high culture, artistic creativity, and aspiration to an ideal (as in the German term for fine arts: *Bildende Künste*) (Pellerin, note 10 in Nietzsche, 2014, p. 50).

3 I acknowledge that the USA has gone to great lengths to form a curriculum that teaches and assesses technological literacy. However, I will discuss later why this appears to be difficult, if not impossible to deliver.

CHAPTER 2

# What Is Meant by Technology, Technique, Technological Knowledge and Ethnotechnological Literacy?

## 1 Introduction

This is an important issue to clarify at the outset, especially when considering education about technology or the concept of ethnotechnological literacy. This clarification is, however, no easy task because the concepts of both technology and technique are complicated and somewhat confusing. This is largely because they have become ubiquitous and so subject to a variety of meanings. Complications arise for several reasons, one of which might be considered in a similar manner to the concept of Schrödinger's cat.[1] The two terms, technology and technique can, not unlike Schrödinger's cat, mean both the same thing and different things at the same time. Unlike Schrödinger's cat, however, this does depend upon geographical location, the relevant historical epoch which is under investigation, and context. For this reason, I will start this chapter by considering several extant perspectives on these varying concepts, covering different locations and different historical periods that, collectively, attempt to define technology and/or technique. Taking the concept of technology as my starting point, I will go on to consider the concept of technique, followed by Mitcham's four concepts of technological knowledge, and finish by exploring the implications of the meaning of those terms for the development of a new concept for the ongoing process of becoming ethnotechnologically literate.

In the article '*Heidegger's Hand* (*Geschlecht 2*),' Derrida deconstructs the German word geschlecht. He reveals a significant problem in translating this German word into English. Depending upon the context, it may translate to mean 'sex,' 'race,' 'species,' 'genus,' 'stock,' 'family,' 'generation,' 'genealogy' or 'community' (Derrida, 1987, p. 162). The context that this word relies upon for definition, is liable to cause confusion at best, and at worst, imply more ominous perspectives, such as racism or sexism. Similarly, the concept of technology, etymologically, has had, and continues to have, a difficult background and development. Part of the problem with the term technology, is not unlike that of the case with geschlecht; that is actually being able to define it in some universally agreed way that will not produce confusion. This is problematic

 DOI:10.1163/9789004537002_002

largely because it can mean a myriad of things, depending upon both the context within which it is set, and the meaning attributed to it by the person using the term. This follows the logic of the local over the universal where, whilst universal definitions may be given for technology, local or cultural variations inevitably emerge. Essentially, the term technology is a portmanteau neologism, a made-up word combing techne and logos, a word that has only relatively recently entered into the mainstream lexicon and even now, continues to defy any semblance of a universally accepted definition. According to the Oxford English Dictionary, its first recorded use was in the seventeenth century, but it was not until the 1930s that it became a keyword in Occidental grammar usage (Oldenziel, 1999, p. 14).

There are many derivations today, but none that offer a clearly articulated concept that all can agree upon. Variations on the term range from the philosophical to every-day usage, such as technics (Stiegler), technique (Ellul), actants (Latour), technology (lots of people), originary technicity (Derrida), technologies of the self (Foucault), technical invention, industrial arts, innovation, mechanical arts and even applied science. It is not just these variations that need clarification, but their local cultural expression through language. The English *technology*, the German *technik*, the French *technique*, the Spanish *técnica* and so on, all help to further muddy the waters. For example, the book by the French philosopher Gilbert Simondon entitled *On the Mode of Existence of Technical Objects* is a translation into English from the original French version entitled *Du mode d'existence des objets techniques*. Notice that techniques in the French translates to technical in English, not technique. Another example, highlighting this confusion, is made manifest in the book written in 1946 by Friedrich Georg Jünger. In its English translation it is entitled *The Perfection of Technology*.[2] In its original German it is entitled *Die Perfektion der Technik* and in the French translation as *La Perfection de la Technique*. This clearly demonstrates that, in English language usage, the concept of technology is differentiated from the Continental concept of technique. Technique, in English usage has more to do with the skills and methods required for achieving any particular goal (Schatzberg, 2018, p. 8). Technique in France or Technic in Germany appear to conflate technology with technique.

> The divergence between English and Continental usage did not exist in the nineteenth century when, for the most part, English, French, and German recognised a common set of meanings for technology. This commonality is apparent in both dictionary definitions and actual usage … [T]he scope of technology in English was largely limited to technical education, and was little used, for example, in the social sciences … By the

> nineteenth century, technology in English, French, and German referred to the description, principles, or teaching of the practical arts and, less often, to technical terminology. In this usage, technology was a field of study (branch of knowledge – subject area), not the objects of study. (Schatzberg, 2006, p. 489)

## 2 So, What Is Technology?

> The word technology appears to have different meanings in academic and every-day speech. Although in the past, the term had a very specific, limited, and unproblematic meaning, this is no longer the case. Persons who employed the term spoke of a 'practical art,' 'the study of the practical arts,' ... Most people spoke directly of machines, tools, factories, industry, crafts, and engineering and did not worry about 'technology' as a distinctive phenomenon. (Winner, 1977/1992, p. 8)

The French philosopher Bernard Stiegler, in his *Technics and Time* (1998), understands technology as "organised inorganic matter" (p. 82), and as "the pursuit of life by means other than life" (p. 17). In other local cultural terms, individuals of note have, over the annals of time, been able to produce alternative meanings for the concept of technology. When, in early twentieth-century North America, the highly influential *The New Republic* magazine asked leading intellectuals of the time to name the books that had shaped their minds, Thorstein Veblen was first on their list (Oldenziel, 1999, p. 43). It was he who had helped revive the term technology by forging a bridge between the nineteenth-century discourse of the industrial arts, and the twentieth-century talk of technology as a joint stock of knowledge and experience held in common by civilised peoples. He nevertheless considered technology to be something represented by disembodied objects, devoid of any human agency and ethically neutral (Oldenziel, 1999, p. 45). Veblen linked technology to the process of social evolution by revealing what he saw as a conflict between business and industry, or modern capitalism and consumerism as being "incompatible with the peaceful development of modern industry for the benefit of the entire community" (Schatzberg, 2006, p. 499).

Given that many other variations on the definition of technology exist and in order to attempt a more succinct definition than has been explored thus far, it will be necessary to explore the derivation of the term historically. This will clear a path towards separating, and thereby reconceptualising the two major terms: technology and technique.

The concept of technology, in its modern sense, derives from the Indo-European root '*tek*' which, according to Porkorny (cited in Roochnik, 1996, p. 19), means "to fit together the woodwork of a woven house," as undertaken by a specialist or expert known as a Tektōn: a woodworker. These Indo-European origins date back some eight and a half thousand years, (Atkinson et al., 2005). The Indo-European derivation translates, over time, into the Greek term 'techne' which, in its pre-Homeric meaning "came to refer to the knowledge or skill [associated with one] who produces something from wood" (Roochnik, 1996, p. 19). In Greek terms techne is commonly translated as "art," "craft" or "skill" (Mitcham, 1994, p. 117). Roochnik, in his book *On Art and Wisdom* (1996) goes on to give a detailed analysis of the work of the pre-Platonic Greek poets such as Homer, Solon, Aeschylus and Isocrates, who each offer various accounts of the term techne in their epic poems. Over time, these poems offer a historic and constantly evolving explanation of the concept of techne. These varying accounts move from a primitive conception towards one that is more complex in nature. The primitive conception situates the subject matter of techne as something that is specific and that is orientated towards some preconceived goal that only an expert can be relied upon to undertake. The expert, according to these accounts, must be certified and, significantly, be "unlike someone who has merely accumulated experience, [the expert] can explain something about [his/her particular skill] to those who wish to learn about it; he can help to transform a layman into a fellow expert" (Roochnik, 1996, p. 20). This last criterion stipulates that techne must be able to be taught by the expert practitioner and consequently, in order that it may be taught precisely, its subject matter must be determinate. In other words, techne is rule-driven in a very precise and determinate way. The term technology, in this sense therefore, involves rationalised and therefore efficient, making or producing "that operates upon the nonhuman material world. These activities are most truly techne that involve the greatest quantitative precision" (Mitcham, 1994, p. 119). Techne, in this sense seems to place more emphasis on the development of skill or technique.

This latter criterion persisted as the other more primitive criteria evolved over time, with new and abridged concepts for defining techne beginning to emerge. Tuchē (chance) for example, is removed from any defining concept of techne, thus implying that techne "is a deliberate application of human intelligence to some part of the world wielding some control over tuchē" (Naussbaum, 1979, p. 95). Other criteria also began to emerge for techne such as: the artefact produced is considered useful and reliable by its users; its function becomes identical with its end; it is very precise and requires "complete mastery of clearly stated basic principles from which the body of knowledge is derived" (Roochnik, 1996, p. 50); it is universal in its application. Terms often associated with techne during this period were "mechane (device, machine) and its variants mechanema (contrivance, machine) and mechanasthia (to contrive, to construct)" (Wheeler, 1997, p. 28). Plato, in his dialogues also argues that techne is "craft," "skill" or "expert" knowledge. He then attempts to add medicine, warfare and mathematics to the growing lexicon of meanings for techne as involving subject matter that is quantitative, fixed and immutable. He thereby regards "doctors, generals, and mathematicians as possessing a techne, meaning that they are knowledgeable about a specific subject matter, can transmit their knowledge in teaching, understand general principles or rules that apply across all instances within their field, and can give a rational account of why their practice succeeds" (Janaway, 2006, p. 396). However, many debates about medicine being a techne, in the terms described above, were to contest this quantitative conception opined by Plato. These debates constellated around the concern that medicine, like rhetoric, could not be considered determinate as was the case for the more craft-based forms of techne: an unwell patient, it was argued, may die despite the ministrations of an expert doctor, or, conversely, may live without ever having consulted one. This led several commentators of the time, like Hippocrates, Isocrates and Aristotle, to argue for an alternative to the very deterministic criteria for the concept of techne that had gone before.

Isocrates, for example, argued vociferously against what he referred to as a 'fixed techne' that could have only hard and fast rules. In his *Against the Sophists; 390 BCE,* he criticised the sophists for claiming that they could directly teach rhetoric in the technical form described above. He argued:

> I am amazed whenever I see these [sophists], who fail to understand that they are applying the paradigm (paradeigma) of a fixed (tetagmenē) techne to a creative process, setting themselves up as instructors of youth. For who except them does not know that, on the one hand, correct spelling is stable and remains unchanged, so that we continually and always use the same letters for the same purpose, whereas, on the other hand, when it comes to speeches (logoi), the situation is exactly the opposite. (Isocrates in Roochnik, 1996, p. 74)

Whilst it may reasonably be argued that spelling, together with words, do change over time and are not necessarily fixed and immutable, Isocrates does, however, highlight an important and emerging distinction; the difference between the paradigm of a 'fixed techne' and that of a creative process such as rhetoric or art. In other words, processes involving creativity must have qualitative dimensions and as such, cannot be a techne in its quantitative sense alone. It is Aristotle who takes this argument forward by agreeing with Plato, his former teacher, that whilst medicine and rhetoric do both qualify as forms of techne, they do so conditionally. He effectively removes the necessity for the concept of techne to result in some quantifiable form of production. He argues instead that:

> It does not belong to medicine to produce health, but only to promote it as much as possible ... it belongs to rhetoric to discover the real and apparent means of persuasion. (Rhetoric, 1355b12 in Roochnik, 1996, p. 80)

In other words, as Roochnik illustrates, these forms of techne are not absolute and immutable but are, rather, compatible with failure and, very significantly, have the capacity to be value-laden (1996, p. 80). Diagnosis can vary according to the value judgements of medical practitioners, just as rhetoric can be used to express opposing political and cultural values.

Aristotle takes this argument even further by offering a distinction between two types of knowledge relating to the concepts of value-neutral and value-laden knowledge. These are techne and phronesis and they are considered by Aristotle to form a distinction between two kinds of activity: making and acting. He proffers that these activities are different by arguing that the state of capacity to make (techne) is different from the reasoned capacity to act (phronesis). "For while production [making] has an end distinct from itself, this could not be so with action, since the end here is acting well itself" (Aristotle, 2000, p. 107 1140b). Dunne (1993) interprets this distinction as follows:

> Techne provides the kind of knowledge possessed by an expert in one of the specialised crafts, a person who understands the principles [...] underlying the production of an object or state of affairs, e.g. a house, a table, a safe journey, or a state of being healthy. Phronesis, on the other hand, characterises a person who knows how to live well. It is acquired and deployed, not in the making of any product separate from oneself but rather in one's actions with one's fellows. It is personal knowledge in that, in the living of one's life, it characterises and expresses the kind of person that one is. (p. 244)

In this sense, phronesis is conceived as being value laden, as in the way one person's perspective or state of reason, interacts with other perspectives. Techne, on the other hand, is characterised by the production of some utilitarian object, one designed to be useful or practical rather than expressive,[3] and so is considered to be value-neutral. Dunne (1993), citing Hintikka, speculates that this fascination with technical rationality, or, in Heidegger's terms, "calculative rationality," is due simply to the fact that "of all human activities the activity of the craftsman is likely to have the most concrete and clearly definable product" (p. 250), or, as Dunne goes on to suggest, that in these terms, "it [craftsmanship] is the activity where, with least ambiguity, success can be achieved and assessed ... in terms of their outcomes" (p. 250). However, Dunne (1993) goes on:

> [t]he locus classicus for this theoretical emphasis in Aristotle's conception of techne is the first chapter of the Metaphysics. There the man of techne is firmly distinguished from the man who works merely from memory or from experience: 'we think that knowledge and understanding [...] belong to techne ... and we suppose technitai [plural for those who apply their techne] to be wiser than men of experience ... because [they] know the why and the cause.' Since he is concerned above all with the forms of things in which resides their intelligibility, the technites aims are not just to deal with particular cases, but to reach 'one universal judgement about similar objects.' (p. 252)

It is in this sense that Aristotle suggests a form of logic to govern reasoning, and he places this form of reasoning within his concept of techne. In this way, Aristotle adds to the previous conceptions of techne a form of reasoning that begins to invoke some degree of ethical judgment based upon a logical analysis. He achieves this by placing a new emphasis on understanding 'the why and the cause' into reasoning associated with techne. He thus, sets out a new way of conceptualising technology that would greatly influence modern philosophers of technology, particularly Heidegger. This Aristotelean thinking foreshadows the modern normative conception of thinking about technology, not in terms of 'can we,' but rather in terms of 'ought we'?

It is important to further note that, in terms of techne, the ancient Greeks would produce something for a specified and actual need. If someone needed a house for shelter, or a ship to sail the sea, the person who was recognised as having expertise in the appropriate domain would be commissioned to apply his techne in that respect. Whilst the experts would legitimately expect some form of renumeration for the application of their techne, they would be quite perplexed with the notion of modern mass production, predicated as it is on producing goods for perceived future needs in the interests of profit (Mitcham, 1994). This is another important distinction between the Greek meaning behind the word techne and its modern incarnation.

Feenberg (2006), moreover, highlights two other important distinctions that emerge and which were important for the ancient Greeks in this respect. The first is the distinction made between the concept of physis and poiesis. For the Greeks, physis translates to nature and poiesis translates to the practical activity of human production. The word techne, for them, is associated with the term poiesis. The second distinction that Feenberg (2006) makes is the distinction between existence and essence. “Existence answers the question of whether something is or is not. Essence answers the question of what the thing is. That it is and what it is, appear to be two independent dimensions of being” (p. 6). Thus, for Feenberg, the Greeks made a clear distinction between the idea of natural and artificial, and between the essence of something, which in the technological world, precedes its existence, whereas in the natural world both essence and existence occur simultaneously. The existence of a tree, for example, occurs at the same time as its essential quality: a tree does not design itself and then produce itself. A technology, on the other hand, is designed prior to its existence; its ‘idea’ or essence precedes its existence in order that we may then know it. However, in ancient Greek ontology, artifacts and nature shared a common ground in that technical making and natural self-production shared the same structure. Techne, in this sense, includes both purpose and meaning. Human beings are not, for the ancient Greeks, “masters of nature but realise its potentialities in bringing a meaningful world to fruition. Our knowledge of that world and our actions in it are not arbitrary but are in some sense the completion of what lies hidden in nature” (Feenberg, 2006, p. 6).

As we move beyond the Greeks, the terms techne and logos begin to disappear from the literature as they become conjoined over time and begin, more and more, to be expressed as technology. However, *en-route* to becoming technology, the Romans took techne hostage and in so doing, changed its meaning, as Schatzberg (2018) so eloquently describes:

> The progeny of techne reached a broader audience through Latin, which translated techne as ars. From ars came art and related terms, such as artful and artificial, along with the concepts of liberal arts and mechanical arts. Not until the nineteenth century did writers reduce art, in most usage, to the much narrower concept of fine art. In effect, this shift in meaning, the narrowing of art to fine art, ended a millennia-old tradition of philosophical discourse about productive knowledge and action. (p. 16)

In this respect, over time, technology, which was once considered an art form, became conceptually separated from what has now become known as ‘fine art.’ This separation serves to distinguish one form of artistic enterprise, the fine arts, as being superior to the technical arts. Herein lies a history that continues to this day, one that situates the concept of technology within a sphere that is inferior to the finer arts, very much along the lines described by Simondon which will be discussed in detail in Chapter 4.

Technology, thus, came for many, to represent the study and systematic knowledge of the practical arts as distinct from the fine arts. In other words, it became associated with the application of knowledge rather than a form of knowledge in and of itself.

The Oxford English Dictionary records the first known use of the word technology as early as 1612, in which it cites one Edward Phillips who defines technology as 'a Description of Arts, especially the Mechanical.' However, it was not until Johann Beckmann, a German scientific author and teacher published (in German) his *Anleitung zur Technologie* (Introduction to Technology) in 1777, whereupon technology "came to be established as a term, a concept, even a field of study." Beckmann expands this to define technology as "… the science which teaches the processing of natural materials, or the knowledge of handicrafts" (Beckmann, 1785, p. 17, author's translation). In this volume he sets technology in opposition to nature. Another later set of volumes by Beckmann which have been translated into English as *History of Inventions, Discoveries and Origins*, was first published in German as *Beiträge zur Geschichte der Erfindungen* in 1780. In this collection he sets out to study a panoply of technological inventions including machines for noting down music, diving bells, sawmills, saddles and, bizarrely, the colour pigment ultramarine. In his footnotes, he refers to the technological history of inventions. It was not until 1829 that Jacob Bigelow, a professor of engineering in Boston, published his *Elements of Technology* and made the claim that technology is the application of science in handicrafts (van der Laan, 2016, p. 5). On the other hand, it was in 1877 that the first publication emerged describing a philosophy of technology. Ernst Kapp penned what is considered by many to be his magnum opus *Grundlinien einer Philosophie der Technik* (Elements of a Philosophy for Technology: On the Evolutionary History of Culture).

Finally, to complete the confusion, the term technology is at odds with its own etymology. According to Schatzberg, "The -ology suffix suggests that technology should refer to an academic field or a system of formal knowledge, a meaning derived from the ancient Greek term logos, as 'reasoned discourse.' However, in present-day usage, technology refers more to things than to ideas, to material practices rather than scholarly discipline" (Schatzberg, 2018, p. 8). This relates to Simondon's concept of the dual phase shift from the unity of the magic mode of existence into two realms, one academic and one technical, as will be discussed later in Chapter 4. In terms of technology education today, this bifurcation epitomises the genesis of what, in educational terms, is now referred to as the academic/vocational divide, where technology education is considered, for the most part, to be a vocationally orientated subject and not academic in nature (see, for example, Dakers, 2007a).

## 3 So, What Is Technique?

European cognates of technique, have had, until recently, more or less the same meaning as technology in English. What is strange "is that cognates of technology also exist in all Continental languages, for example die technologie in German and la technologie in French" (Schatzberg, 2018, p. 8). Whilst technology cognates were known in Continental language, they were, however used much less frequently than cognates of technique. Cognates of both the terms technology and technique translated into English as the singular term 'technology.' Indeed, Schatzberg goes on to suggest that, etymologically, technology is the science of technique. Technique he maintains, refers to the principles and processes of the useful arts, which have more or less the same meaning as the Greek term techne. Significantly, however, he also points out another primary meaning for technique as skills and methods for achieving any goal. It is this second meaning that is the closest match for the English technique (Schatzberg, 2018, p. 9). So, the term technology, in common parlance, appears to be perceived of as a noun. Technique, on the other hand, is perceived as being a process, a learned physical movement of the body, evolving over time from novice to skilled action, resulting in something being produced: learning to sing; to play a musical instrument; to use a technology; to design something, etc.

Citing from Max Weber's *Economy and Society*, Gertrud Koch outlines his theory of technique as instances of an action that are part of a wider context into which they merge. In other words, the concept of technique, for Weber, can only ever be understood in terms of context.

> The "technique" of an action refers to the means employed as opposed to the meaning or end to which the action is, in the last analysis, oriented. "Rational" technique is a choice of means which is consciously and systematically oriented to the experience and reflection of the actor, which consists, at the highest level of rationality, in scientific knowledge. What is concretely to be treated as a "technique" is thus variable. The ultimate meaning of a concrete act may, seen in the total context of action, be of a "technical" order; that is, it may be significant only as a means in this broader context. Then the "meaning" of the concrete act (viewed

> from the larger context) lies in its technical function; and conversely, the means which are applied in order to accomplish this are its "techniques." In this sense there are techniques of every conceivable type of action, techniques of prayer, of asceticism, of thought and research, of memorising, or education, or exercising political or hierocratic domination,[4] of administration, of making love, of making war, of musical performances, of sculpture and painting, or arriving at legal decisions. All of these are capable of the widest variation in degree of rationality. (Weber in Koch, 2019, pp. 8, 9)

Ellul, in his 1964 book; *The Technological Society*, which, in its original 1954, French form, was *La Technique ou l'enjeu du siècle*, considers technology and technique as noun and verb consecutively. As an example, Ellul differentiates the technological or artificial world from the natural world. He uses the example of large, urbanised conurbations to demonstrate the way human beings are changing the natural world-order on a grand, and in his view, unsustainable way. He concludes that it is the application of technique and the consequential technological artifacts that follow, that are proving to be an ongoing problem for the sustainable future of humanity. He argues that technique, in this respect, results in the exploitation of nature as humanity's hunger for new and ever more efficient technologies grows exponentially, thereby adversely affecting humanity's moral and ethical compass. Technique, for Ellul is "the totality of methods rationally arrived at and having absolute efficiency (for a given stage of development) in every field of human activity" (Ellul, 1964, p. XXV). 'Know how' flips around to 'means justifying the ends.' What is useful takes precedence over what is harmful, physis is sacrificed to the technique of poiesis, and in so doing, humanity loses all perspective on how to live harmoniously with nature. In Ellul's conception, "life is not happy in a civilisation dominated by technique" (Ellul, 1964, p. VIII).

What starts to become evident from Ellul's account then, is that technique is related to the process of human activity in all things. Technique is something specific, orientated towards some preconceived goal that involves rationalised, and therefore efficient, making or producing no matter what the cost. Or, as Mitcham states, it is quantitative precision "that operates upon the nonhuman material world" (Mitcham, 1994, p. 119). Ellul's concern is that as human beings evolve, so too do their technical capabilities. In a technological society, human applied techniques evolve through the apparatus of evermore sophisticated inventions, systems that can only now be achieved by the ruthless exploitation of the planet without any serious concerns over the cost. Ellul, to all intents and purposes, was prophesising global warming in the Anthropocene.

In more general terms, the Oxford English Dictionary Supplement defines technique as: (a) Now used in the manner of execution or performance in any discipline, profession, or sport, or a skill or ability in any of these; (b) in the sense 'technical or artistic skill' ... and (c) loosely, a skilful or efficient means of achieving a purpose; a characteristic way of proceeding; a knack (OED, 1987).

Technique, in this sense, accords with David Pye's (1999) definition. "A technique embraces a group of systems which cause some specific change in the characteristics of pieces of material – any material" (p. 43). Pye then goes on to list several processing techniques which cause change to occur on material:

> Processing material means altering its properties, as in smelting or in hardening steel by heat, or making powder out of a solid, or cement out of mud. Wasting techniques are those which involve carving away a piece of material until the shape you want remains. Sawing, milling, planing, turning, punching, are examples. Construction means making a whole out of parts, by connecting them. Connection may be by interpenetrating, as in a carpenter's joints or a rivetted joint, or by interlocking, as in spinning or weaving, or by welding, or by jointing with an adhesive. Forming means changing the shape of a piece by bending, pressing, forging, etc. Casting means pouring a liquid into or over a mould. (p. 44)

The definition of technique as espoused by Pye, is related to the creation of technical objects. However, go into any bookstore and you will find, in keeping with the *Oxford English Dictionary* definition, books outlining techniques associated with playing musical instruments, techniques related to cooking, techniques involved in sport, sexual techniques, singing techniques and so on. Techniques in these areas are skill related. They are procedural in nature and are learned by doing and perfected through repetition.

Learning and teaching new skills or techniques, therefore, is a process where a learner begins at the level of novice and (may) ultimately progress to a level of expert or somewhere along that continuum (Dreyfus & Dreyfus, 1986). Learning techniques associated with craft-based technology education, based on this model, begins by breaking down tasks into what can be figuratively described as 'bite sized pieces.' Essentially, tasks are broken down into simpler, context free elements or parts. These parts are based upon rigid adherence to taught rules or plans, have little situational perspective, and allow for no discretionary judgement. At this level the teacher supplies the facts and procedures which the pupil needs to learn. The teacher demonstrates to the class the psychomotor skills involving tool use. Great emphasis is placed upon the correct procedures associated with that tool, including safety aspects which play a crucially important role. The pupils copy, within the confines of their ability range at the time, the techniques demonstrated by the teacher. Repetition by the novice plays an important role within this domain. Whilst a degree of creative flexibility and rule breaking should be encouraged in what might be loosely described as the design stage, rules governing the design process may begin to be introduced even at this stage.

At the opposite end of the continuum from novice, is the expert:

> The *proficient performer*, immersed in the world of his skilful activity, *sees* what needs to be done, but has to *decide* how to do it. The expert not only sees what needs to be achieved; thanks to his vast repertoire of situational discriminations, he also sees immediately how to achieve his goal. Thus, the ability to make more subtle and refined discriminations is what distinguishes the expert from the proficient performer. (Dreyfus, 2001, p. 41, original emphasis)

Expert level is thus seen to be at a level which is more in keeping with a well-established craftsman or designer. To achieve expertise, the performer must be a specialist practicing within a specific domain with a vast amount of experience in that domain. This level of skill development is clearly beyond the scope of school craft-based technology education.

> An expert generally knows what to do based on mature and practiced understanding ... An expert's skill has become so much part of him that he need be no more aware of it than he is of his own body ... the expert business manager, surgeon, nurse, lawyer, or teacher is totally engaged in skillful performance. *When things are proceeding normally, experts don't solve problems and don't make decisions; they do what normally works.* (Dreyfus & Dreyfus, 1986, p. 111, original emphasis)

Expertise suggests a large degree of immersion in the specific subject domain and less of a generalist approach. Evidence suggests that for skills to be acquired at this level, authentic practise contexts are required (Hager & Hyland, 2003; Dreyfus, 2001). It is extremely unlikely that any student studying craft-based technology education can acquire the skills and knowledge required to reach the level of expertise described above. School-based learning within a craft-based, vocational orientation can, therefore, only ever be synthetic in nature. To learn a skill in a work-place situation will always be a different experience from learning within a school. Work comes with a different set of responsibilities that are not present in school. For example, what you do in the workplace must satisfy your employer and your income will be reflected in that. The closer you become to being expert, the more employable you become. This is not a pressure that exists in school. This also applies to designing in technology education. It is a safe environment within which to experiment, if pupils are given the opportunity. This questions any claims that the subject is, or should be vocational.

It can be argued that this model of skill-based learning, when applied, is necessary in workshop environments. It is clearly predicated upon the development of procedural (knowing how) and declarative (knowing that) models of pedagogy. As was discussed above, to reach a stage of expert competency, the expert must also be able to teach others the relevant skills and to do so, the repeated accurate application of these skills must "involve the greatest quantitative precision" (Mitcham, 1994, p. 119). This model of pedagogy is, however, antithetical to the ongoing process of becoming ethnotechnologically literate. "Understanding a language or a science is much more than memorising the elements and the rules relating to them. The student needs not only the facts but also an understanding of the context in which that information makes sense" (Dreyfus, 2001, p. 34). I shall put my concept of ethnotechnological literacy into context momentarily, but first I want to explore the concept of technological knowledge.

## 4 Carl Mitcham's Four Concepts of Knowledge Relating to Technology and Technique

Carl Mitcham in his seminal book *Thinking Through Technology* (1994), develops his four forms of knowledge related to technology. He suggests that the concept of technological knowledge, like the concept of technology and technique, is a complex one. He suggests that the concept of technological knowledge goes beyond the application of knowledge alone. He contrasts knowledge about nature with knowledge about technology, where aspects of technological knowledge are required for designing, fabricating, and using technical objects (Mitcham, 1994). Knowledge about nature, on the other hand, was more about the science of discovery until genetics and other similar interventions redefined it as technoscience, in which aspects of nature are subject to technical manipulation. Technological knowledge is both procedural (knowing how) and declarative (knowing that). When engaging with tools and machines in technology education settings, for example, knowing that such and such a tool or machine has a specific use and a unique identity, is to know it declaratively, whereas to know how to use it, is to know it procedurally. This tends to be the dominant orthodoxy in craft-based technology education workshops, laboratories, and graphic drawing classrooms. However, Mitcham adds to the concept of technological knowledge, three other forms of knowledge related to technology: technology as object, technology as activity and technology as volition.

For the concept of technology as object, which falls under the rubric of technology and not technique, Mitcham discusses technical objects

> in a broad way to include technological systems, as a series of objects. He further describes how technological objects could be either static (structures, such as buildings or roads) or dynamic (machines, such as cars or computers). A technological object can also be divided into networks and processes [a component in a machine for example] ... This is the most common way people describe technology. (Svenningsson, 2019, p. 15)

This type of knowledge is more declarative in nature.

Technology as activity is, for the most part, related to technique which, according to (Mitcham, 1994), forms the relationship between humans and technological devices, such as being able to operate machines or maintain technological systems. It can also relate to designing and constructing technologies. Once again, declarative and procedural knowledge constitutes a major component of technological activity. However, when applied to the activity of inventing or designing, technological knowledge, in combination with technological activity, becomes more conceptual in its application: 'if I do this, the result might be some novel change.' This can also relate to Ellul's concerns discussed above: 'if I do this, the result might be some novel change,' with the addition, as part of the discussion, of: 'will whatever change results be sustainable or detrimental to the planet?' These latter types of questions, I argue, do not appear to feature large in craft-based technology education today.

Finally, technological volition is more related to the philosophy of technology and includes areas such as ethics, morals, environmental and sustainability issues, as well as political issues. The concept of technology as volition thus serves to transform knowledge about technology in objective terms into new ways of thinking about technology subjectively. This involves much broader conceptualisations of technology, more critical and complex ways of thinking about the human/technology interrelationship and how that can influence both culture, society and the environment.

Considering the concepts of technology and technique as discussed above, leads me to certain conclusions that are germane to this book. When taken from a pedagogical standpoint, I find two schools begin to emerge. The present model of craft-based technology education, mostly defined by the learning of specialised craft skills and/or techniques that are closely associated with engineering and product design, forms one school. The other model, the one that I am proposing, is more associated with technological volition as described by Mitcham above, or technological literacy as proposed by me (Dakers, 2014a, 2014b). The former is, for the most part, predicated upon the development of procedural and declarative forms of knowledge. This is no bad thing per se. There are many advocates for the learning of psychomotor skills and the development of craft skills and knowledge through tool use. However, this technology education paradigm lacks any possibility of its incorporation into the ongoing process of becoming ethnotechnologically literate, which I will go on to define below. This may be for a number of reasons. It is, for example, difficult, if not impossible, to formally assess issues that are incorporated in the affective domain, such as ethics, morals, environmental and sustainability issues, as well as political issues. These are based upon individual values which cannot be assessed in terms of right or wrong, although they can be considered to be right or wrong if they are in direct opposition to the values held by another individual. This can be exemplified in the conflicting debates surrounding the Covid 19 virus, such as the debate between vaccine sceptics and those who supported vaccination.

It is for reasons such as these that I am advocating a new pedagogy, one which lays much less emphasis on the development of procedural and declarative knowledge; one which results instead in a progressive pedagogy that prioritises the development of conceptual forms of, not so much knowledge acquisition, but personal knowledge development that enables student voice and that encourages students to actively participate in debates regarding the technologically textured world we all inhabit. As Giroux advocates, in terms of education in general, but I would argue equally in terms of education about technology:

> [Technology] education is fundamental to democracy and .... no democratic society can survive without a formative culture shaped by pedagogical practices capable of creating the conditions for producing citizens who are critical, self-reflective, knowledgeable, and willing to make moral [and ethical] judgements and act in a socially responsible way. (Giroux, 2020, p. 1)

This distinction between the more philosophically orientated concept of volition and other forms of technological knowledge as espoused by Mitcham above, follow essentially the same distinctions highlighted by Aristotle's difference between the capacity to make (techne) and the reasoned capacity to act (phronesis), Feenberg's distinction between nature (physis) and the practical activity of human production (poiesis), Schatzberg's distinction between technology as referring more to things than to ideas, to material practices rather than scholarly discipline, Ellul's fears for the planet where the artificial (technological) is exploiting, and so wreaking havoc with nature. This also relates to Simondon's concept of the dual phase shift from the unity of the magic mode of existence into two realms, one academic and one technical, which will be discussed later.

## 5 My Translation of Technological Literacy into Becoming Ethnotechnological Literacy

Over the years I have been attempting to define technological literacy. My first book in 2006 entitled *Defining Technological Literacy* offered a myriad of perspectives from diverse fields such as philosophy, sociology, anthropology, psychology, education and more. I followed up on this same field of enquiry by bringing together new thinkers in 2014, with a book entitled *New Frontiers in Technological Literacy*. My chapter entitled, *Technological Literacy as a Creative Process of Becoming Other*, although written some 9 years ago, served as the original nucleation of the crystal that, as it has grown, has led me to this place in this time. I have concluded that technological literacy is not the same as craft-based technological education. Technological literacy is more concerned with the ethical, sociological and cultural aspects of the technologically textured world we human beings occupy. Craft-based technological education is more concerned with the material aspects of the technological world, made manifest principally in the fabrication of products and sometimes in design. However, they both use the same adjective, 'technological' to describe each noun,[5] 'literacy' and 'education.' This implies that technological literacy is part of craft-based technological education, which it is not, as I will argue in a later chapter. I have, therefore, combined 'ethno' with 'technological' to make this distinction. My inspiration for this came from a book written in 1972, entitled *Technology and Social Change*, a book about social anthropology that recognised a growing interest, at the time, in 'ethnotechnology.' This had the aim of contextualising the relationship between technological development and other aspects of culture and society. The term *ethno* has its derivation in *ethnos* which comes from the Greek, meaning people, nation, class or tribe. It is related to ethnicity, all of which combine to depict culture or society. The editors close their book with the following observation:-

> Work is needed to provide the 'contextualisation' for assessing the relationships of technical development to other aspects of culture and society. To throw light on the human side of technological change, we must study the cognitive, 'ethnotechnological,' and psychological dimensions of individual projects and products. Too frequently it appears that technological studies have been concerned with profit-and-loss economics and 'modernisation' from the outsider's point of view, rather than with the interpretations used by the affected populations to define their situations and formulate adaptive strategies. It has been the special concern of anthropologists to explore that 'actors point of view' in situations of sociocultural change. This perspective, if used in judicious combination with effective [qualitative] observations, can add the vital humanistic dimensions to technological studies. (Russell & Pertti, 1972, p. 331)

I very much agree with their summary and have developed the concept to coin the term ethnotechnological literacy. I believe this will help distinguish the difference between this concept and craft-based technology education. I do admit, however, to having substituted qualitative for quantitative in the above citation because, as will become evident as this book progresses, the qualitative dimension is much more closely related to social and cultural ethnotechnological literacy, whereas the quantitative dimension is more aligned to craft-based technology education.

The concept of technology, technique and technological knowledge, as discussed above, demonstrates that to understand those terms, context is vital. This applies equally to the concept of technological literacy. There have been many variations offered over the years as to what constitutes technological literacy, some having common elements, others being more distinct. Todd sums it up succinctly: "Technological literacy is a term of little meaning and many meanings" (Todd, 1991).

Probably the largest set of volumes ever produced relating to technological literacy in school settings have been the ITEEA[6] Standards for technological literacy in the United States. These standards, which include a volume on assessing technological literacy, form an ongoing process in the USA. They are currently going through a review process which now challenges some of the older definitions (Buelin et al., 2019).

I do not intend to list the plethora of definitions offered thus far, suffice to say that the definition I will offer here is radically different. Given the many, varied, ongoing definitions of what constitutes technological literacy, I will offer my definition; one that has been informed by the discussion at the beginning of this section, together with the philosophies of those philosophers mentioned earlier. I will not only discuss my version in this chapter but will continue to refer to it throughout the book. However, I do want, initially, to look at a change in emphasis being given to technology education and how that will factor into my new definition.

There appears to be a paradigm shift occurring in what was once designated 'technological literacy.' This shift in emphasis is being made manifest now, as STEM literacy. The Design and Technology Association (DATA) in the UK, suggested in 2014 that their rationale for studying technology education was to prepare;

> young people to cope in a rapidly changing technological world; enable them to look to think and intervene to improve that world; develop skills required to participate responsibly in home, school and community life, help students to become discriminating consumers and users of products ... encourage the ability to consider critically the uses, effects and values dimensions of design and technology. (DATA in Michael, 2014, p. 48)

The DATA website today, however, offers a new and revised rationale:

> Design and Technology is about providing opportunities for students to develop their capability, combining their designing and making skills with knowledge and understanding in order create quality products. (DATA, 2022)

A search on the DATA website for STEM, moreover, reveals 117 results. A search for "technological literacy" reveals 0. The same can be said for the American ITEEA website (see Michael, 2014, note 4). It is for this reason, together with those that will follow, that I feel compelled to reconceptualise technological literacy as ethnotechnological literacy, a title that clearly indicates a subject matter that moves beyond the mastery of technique or the assimilation of objective technical knowledge. Rather it embraces a concern with substantive values, values that are not set out explicitly but are rather, presented abstractly as, for instance, a commitment to improvement of the world, to criticism, to creativity and to autonomy.

Ethnotechnological literacy, for me, is not something that that can be taught *per se*, and certainly not in terms of the transference of knowledge from expert to novice. Ethnotechnological literacy, for me, is more related to Aristotle's concept of Phronesis, which considers technology in terms of values, one in which recognition is given to the values and perspectives held by all individuals, relative to the way in which technology affects each of them uniquely in the world they occupy. Conventional craft-based technology education inevitably involves the production of something tangible that is separate from the individual: for example, a pencil case, a working electric circuit or even a design portfolio that follows a preordained process in its production. Often in the early years of secondary, craft-based technology education, it is the teacher who designs the artifact that is to be created or who selects the skills to be learned. These in turn are specified by the curriculum, such as, for example, knowing how to draw a truncated cone. The whole class observes the teacher's demonstration of the drawing and then each student replicates it. The work produced by each student is then assessed by the teacher, who allocates a grade. From the students' perspective, the whole process is designed to enable the acquisition of craft-based skills and techniques. They have no say in what they are producing. This exemplifies a value-neutral perspective of technology, the only thing of value being the grade awarded. This can serve to motivate or demotivate the student, according to their personal experience of, and interest in the process. When I was a technology teacher, for example, I remember one student telling me that his two older brothers had, some years earlier, made a C.D. rack identical to the one he was now making. He had, consequently, little interest in the completed artifact. The ongoing process of becoming ethnotechnologically literate, on the other hand, has no end-product separate from itself. It is, as Dunne (1993) proffered above, "personal knowledge in that, in the living of one's life, it characterises and expresses the kind of person that one is" (p. 244).

In this sense, phronesis is conceived as being value-laden in the way one person's perspective or state of reason, interacts with other perspectives. The students' values involved in living in the technologically textured world they themselves occupy, are given expression and respected, providing that they are expressed in a reasoned and considerate manner; one that is respectful of any opposing values expressed by others. Techne on the other hand, is characterised by the production of some utilitarian object, one designed to be useful or practical rather than expressive, and so is considered to be value-neutral. It is possible for a product to be expressive, but that is usually limited to the later secondary stages if design is incorporated into the subject.

Learning, moreover, is socially constructed and so learned through experience, all experience, not merely that encountered in school. Vygotsky (1978) argues that formal learning must integrate with what he calls spontaneous or tacit learning in a sociocultural setting. He espouses a model where the social precedes the psychological in ontogenesis. This view subscribes to the notion that we are all social beings first, and only then develop as individuals by way of a process of internalisation, subsequent to sociocultural inheritance. In Vygosky's frequently quoted law of cultural development:

> *An interpersonal process is transformed into an intrapersonal one.* Every function in in the child's cultural development appears twice: first, on the social level, and later, on the individual level; first *between* people (*interpsychological*), and then *inside* the child (*intrapsychological*). This applies equally to voluntary attention, to logical memory, and to the formation of concepts. All the higher functions originate as actual relations between human individuals. (Vygotsky, 1978, p. 57, original emphasis)

However, Vygotsky's theory of cultural development does not simply occur in a series of independent single events that go from being between people to becoming inside the child, or adult for that matter. It is an ongoing process of many transformations and constant re-evaluations leading eventually to the gradual emergence of particular perspectives. And these perspectives will differ depending upon the particular social milieu that the child occupies at any given time. An admittedly extreme example might serve to illustrate this. A young person brought up in a social milieu advocating white supremacy, is more likely to develop a very different set of values than those of a young person brought up in the social milieu of a black community.

Vygotsky's concept of development (technological or otherwise), centres around the mediation of socially distributed signs, external to the learner, which take the form of tools which are both semiotic and artefactual in nature, and which come together to form a representation of the culture and society from which the individual is emerging. However, as we will see later, social milieu's are made manifest in many and varied ways during an individual's life, and cannot be predicted in advance. Unforeseen circumstances dictate the trajectory of our lives. Given that humans have always needed technologies for their existence, and that those technologies can only ever exist because of humans, it is self-evident that this symbiotic relationship both effects and affects the way human beings, in association with their technologies, impact upon each other and upon the world. It is as a result of this human/technology relationship that Wiebe Bijker, Thomas Hughes and Trevor Pinch introduced the concept of the *Social Construction of Technological Systems*, also known by the acronym SCOT (Bijker et al., 2012).

Ethnotechnological literacy, in my definition, is thus less about Mitcham's technological knowledge as a requirement to use and make technologies, nor is it about knowing technology as object and only marginally about technology as activity. It is fundamentally about technology as volition or what Nietzsche refers to as the will to power. This relates to human choice about the way technology is utilised, which, significantly and importantly, will be expressed, as Vygotsky proffers, from a multitude of differing perspectives. In terms of technology education in particular, Keirl (2015) argues that the phenomenon of technology means that:

> we humans cannot "be" without Technology and Technology "is" by human intention and (inter)action. That is, technologies and humans coexist intimately. However, despite this intimate human-technology phenomenon, there is a huge irony: why, then, do we not have a parallel education to help understand the phenomenon? Of course, we do have Technology Education and its variants around the world and we might argue that it is not their place to wrestle with the complexities of the phenomenon itself but, rather, they should/can only attend to local matters in a timely and manageable way. However, on closer examination, we can see that the phenomenon and practice in schools are not readily separated – or if they are, only a partial technology education may be taking place (and I use the term "partial" here in both senses: of being limited and being biased). We soon see that the phenomenon and the irony both beg special attention through curriculum and teaching. (p. 14)

My definition of ethnotechnological literacy constitutes a major paradigm shift away from extant models of craft-based technology education as delivered around the world. These differences are made manifest in both content and pedagogy. There has never truly existed a place in these extant models for the delivery of (ethno)technological literacy, despite many attempts. There is certainly no place for my version in most schools, unless some major rethinking is given to the purpose of school education overall, and in-particular to the pedagogies employed. This is not new thinking; progressive changes such as those that I advocate, have been collecting dust in bookcases for the last hundred years at least, but are constantly ignored by successive governments around the world, just as climate change is being ignored in modern times. However, change can only ever come about beginning with new ideas and that is what I am offering in this book.

In this chapter I have demonstrated that the concept of technology and technique, although ubiquitous in today's culture, remains subject to a variety of definitions which tend to be context dependent. This, together with an understanding of the derivation of the concept of technology, forms an important role in the ongoing process of becoming ethnotechnologically literate, as does Mitcham's concept of technology as volition. This chapter has thus informed my definition of ethnotechnological literacy as something distinct from craft-based technology education. In my definition, I use the term becoming ethnotechnologically literate to denote an ongoing never-ending process, rather than a state that one finally reaches. This is a concept that I will explore further in relation to the philosophy of Deleuze.

## Notes

1 A cat is placed in a steel box along with a Geiger counter, a vial of poison, a hammer positioned over the vial, and a radioactive substance. When the radioactive substance decays, the Geiger detects it and triggers the hammer to release the poison, which subsequently kills the cat. The radioactive decay is a random process, and there is no way to predict when it will happen. Physicists say the atom exists in a state known as a superposition – both decayed and not decayed at the same time. Until the box is opened, an observer doesn't know whether the cat is alive or dead – because the cat's fate is intrinsically tied to whether or not the atom has decayed and the cat would, as Schrödinger put it, be "living and dead … in equal parts" until it is observed (Kramer, 2013).

2 The perfection of technology was originally entitled 'The Failure of Technology: Perfection without Purpose.'

3 Technological objects can be expressive, however as techne, more emphasis was given to utility.

4 Relating to governance by priests or other religious ministers (The Chambers Dictionary, 2003).

5 I am describing this in the English language. I accept that the grammatical nuances may differ across language and cultures.

6 ITEEA is the International Technology Education and Engineering Association whose headquarters are situated in Reston Virginia, USA.

CHAPTER 3

# What Are the Characteristics of (Much) of Current Technology Education That Causes It to Fail in the Delivery of Ethnotechnological Literacy?

## 1 Introduction

In the preceding chapter, I argued that the ongoing process of becoming ethnotechnologically literate has more to do with the qualitative aspects of technology as volition, aspects that are socio-culturally orientated and thus subjective. It is, therefore, impossible to measure objectively by summative examination alone. This is because subjective perspectives are immanent to the individual, resulting in the creation of different worldviews. Such literacy can consequently be properly assessed only by utilising formative protocols, a methodology that appears to make policy makers uncomfortable. In contrast, craft-based technology education is more quantitative in nature and so objective. It can be measured and assessed easily by summative assessment alone. I argue, therefore, that for that reason alone, ethnotechnological literacy cannot be integrated into, or mapped onto, extant models of craft-based technology education. Therefore, although the development of (ethno) technological literacy is, in theory, generally considered to be an important aspect of technology education, as is evidenced by curriculum documentation around the world, it consistently proves, in reality, difficult, if not impossible to achieve. Even if policy makers were to insist that ethnotechnological literacy be taught, they would no doubt insist that the learning outcomes be entered into specified rule driven curriculum documentation together with formal summative assessment protocols. This would at once push it into the quantitative realm, transforming it into information to be taught and later recalled. The reality, at present is that the concept of technological literacy, whilst regarded as an essential component of technology education continues to be considered as an add-on, an unimportant addition to be rehearsed before any examination or simply ignored. This is especially the case if practicing teachers and student teachers are not given support in understanding both what the concept of (ethno)technological literacy is and developing a deep understanding of formative assessment protocols.

It is also important to be able to differentiate between a quantitative pedagogy which involves knowing how and knowing that, and a qualitative pedagogy involving volition. In other words, given that technology has always had

 | DOI:10.1163/9789004537002_003

an ongoing relationship with human beings and has evolved alongside them, it is evident that ongoing phase changes occur, resulting in the never-ending ongoing process of socio-cultural evolution. In the delivery of the ongoing process of becoming ethnotechnologically literate, a qualitative form of pedagogy must take precedence over the quantitative pedagogy that forms the bedrock of craft-based technology education today. This chapter will demonstrate that the ongoing process of becoming ethnotechnologically literate is not something that can be finally attained and examined summatively. To that end, it is highly unlikely that it can articulate with craft-based technology education.

In 2006, technology education was thought to be in a somewhat perilous state and in need of an "axiological turn" (Kroes & Meijers, in de Vries, 2018a, p. 15). Governments around the world were questioning the relevance of technology education in their national curricula and felt that there was a need to modernise the subject by bringing it into line with the modern world. Consideration must be given, they thought, as to how technology impacted upon societies and the environment. It was from these considerations that the concept of technological literacy was born. Concerns were also expressed about the declining number of hours allocated to the subject as well as the decreasing number of available teachers. Perhaps the most concerning issue raised, however, was the reduction in budgets for the subject area and for technology education research. These concerns were expressed in the *International Handbook of Technology Education: Reviewing the Past Twenty Years*, edited by Marc J. de Vries and Ilja Mottier, and published in 2006 by Sense Publishers. To be fair, all was not completely dystopic. New Zealand, for example, was at the time working collaboratively with all stakeholders, including the government, to introduce a much more progressive technology education curriculum including the potential integration of technological literacy.

Thinking about ways to proceed in the light of the concerns expressed above, de Vries concluded that in order to build up a sound technology educational theory and practice:

> We need to work on a technology education practice that convinces governments, school administrators and school boards, teachers, parents, and pupils that teaching and learning about technology is not only fun but also a necessary component of every (future) citizen's general education. (de Vries, 2006, p. 9)

That was 16 years ago and de Vries, as editor, was expressing the views that were expressed in many chapters of the review.

Moving forward 12 years to 2018, had anything in fact changed? Unfortunately, not so much. As editor of the Springer *International Handbook of Technology Education*, de Vries, once again cast a dark shadow over the perceived status of technology education around the world. As before, status, funding and resources feature as perennial problem areas and even the new progressive curriculum that New Zealand had been introducing had, by that stage, lost impetus due to an inability to convince the government of the need for change. As de Vries lamented in his chapter entitled 'Technology Education: As International History,' "[h]opefully in due time, there will be a second *International Handbook of Technology Education* with a new chapter in the history of technology education that will show that technology education has been able to overcome the hurdles of survival and flourishes in many countries worldwide" (de Vries, 2018a, p. 83). Keirl, moreover, pinpoints the problem with progressing technology education as being down to policy makers and in particular to successive governments' lack of continuity. He argues that technology education:

> arguably has the potential to redesign itself to move to center, rather than being at the periphery, of a much-needed global education, but this cannot happen while it remains servant to the epistemological and political interests of a few. In fact, precisely because [technology education] has some duty to address both the human-technology phenomenon and the polypotency and multistability of technologies, it can do so as a field grounded in advancing the common good – ethically, sustainably, and democratically. In such ways, aims-talk becomes a prerequisite to a flourishing, if complicated, curriculum for [technology education]. (Keirl, 2018, p. 223)

I will show, as this chapter progresses, more contemporaneous accounts of much the same set of problems, certainly in terms of the USA and England. However, whilst I support these calls for new thinking, as well as funding and support for technology education, my interest in this book is limited to learning about technology in the form of the ongoing process of becoming ethnotechnologically literate. To that end, as this chapter title makes clear, I will argue that craft-based technology education in its current format, is not capable of delivering ethnotechnological literacy, as I see it, for several reasons.

## 2 The Old Is Dying and the New Cannot Be Born

Although Gramsci (1971, p. 371) was not talking specifically about technology education, I have headed this section with this quotation because I feel it to be relevant. Despite a significant number of attempts over the years to reform what I regard to be an outdated technology education paradigm, the subject remains stubbornly rooted in its craft-orientated pedagogy. This is despite much research on the need for change. I shall in this section offer a selection of ongoing evidence to support my argument, which is, as Gramsci elucidates above, a subject that is dying for a variety of reasons but continues to resist being born again anew.

## 3 Conceptual Issues That Are Lacking in Craft-Based Technology Education

It is Francis Bacon (1561–1790) who coined the term 'knowledge is power.' It is also he who attempted to universalise all knowledge into a form that became the foundation of modern science, a subject domain that, he claimed, would discover everything about the world, scientifically, by means of empirical proofs. To be included in these modern scientific practices, were the local knowledges relating to the various crafts employed at the time. This was to be done, according to Bacon, in order to formulate universal, scientifically proven methods applicable to all forms of knowledges and techniques (Bacon, 2019).

Deleuze and Guattari make a differentiation between what they call 'royal science' and 'nomad science' (Deleuze & Guattari, 2004, pp. 361–374). Whilst the former is related to the knowledge of irrefutable facts, enabling those in the know to develop 'Baconian' power, the latter relates more to local and individualised knowledge. These are forms of knowledge that are more malleable and empowering in nature, including the knowledge and skills of the craftsman. Bacon wanted to extrapolate this local knowledge and turn it into scientific knowledge or royal science. Holland explains it thus:

> Owing to the power of royal science to extract abstract concepts from the concrete operations of productive practice, conception and execution become distinct activities, and each gets assigned to a distinct status group. Francis Bacon's program for the development of early modern science illustrates this process perfectly: he charged agents of the Royal Academy with the task of visiting local workshops to extract whatever knowledges were in practice there, and then bringing them back to the academy, where they would be elaborated into formal scientific knowledge, only to be eventually reapplied to the production process in the form of technology, thereby liquidating the autonomy of the workers and subjecting them to technicomanagerial control. (Holland, 2011, p. 19)

Marx takes up this theme in his description of the de-skilling of the various artisans who were eventually brought into the factory system, but without doubt, the Baconian scientific revolution laid the foundations for the birth of the Industrial Revolution, and in more 'modern' conceptualisations, for Taylorism and Fordism. The vocationalisation of craft-based technology has its genesis in the industrial revolution.

## 4 The Industrial Revolution

The pioneers of the industrial revolution were very much guided by Adam Smith who, at the time in his seminal work *The Wealth of Nations*, presented "… a picture of society as an efficient system in which educational institutions play a crucial role, justified in terms of their utility" (Standish, 2003, p. 224). Smith argued that economic progress in this new 'civilised' society was dependent upon the division of labour. The formation of public schooling enabled the retention of class divisions in which those from higher social status attended private schooling whilst those from the lower ranks received a state education that was not so much intended for the private consumption of the learner but contributed rather to the "formation of a particular habitus. It prepared persons for their position in the social hierarchy of ranks" (Vanderstraeten, 2004, p. 196). This social ranking has its roots in Aristotle and Plato who advocated a "hierarchy of theory/practice/production [which] was not only epistemological, but also social in that a person's place in the city state reflected the kind of knowledge that was their daily concern" (Hager & Hyland, 2003, p. 272). "Early in the [18th] century it began to dawn on the administrators that increasingly sophisticated technology would work unprofitably unless some efforts were made to train more workers in factory skills" (Burke & Ornstein, 1997, p. 211). Clearly, this utilitarian and vocationally oriented education had less to do with altruism on the part of industry but was, rather, more inclined towards increasing productivity and efficiency in the workplace.

In the UK, however, this form of educational instrumentality, together with the resulting divisions of labour that emerged, eventually led the "working classes" to express a desire for a more holistic education that went beyond that of industry's immediate needs. This led to the establishment of mechanics institutes that offered evening classes.[1]

Following on from this, and largely as a result of it, a strong movement towards the establishment of "working class" education began to take shape. Funding from the government, and more particularly the trades unions, resulted in the provision of extra-mural evening classes, or continuation classes as they were known. The raison d'être behind this egalitarian philosophy was very much an intention to enhance "an intellectual understanding, not just technical dexterity" (Paterson, 2003, p. 92). Indeed, Heriot Watt College (forerunner to Heriot Watt University in Edinburgh), although offering vocational courses at the time, made it explicitly known that it sought to attract students who wished to "... rise above the position of being mere machines performing certain mechanical and routine operations, in order to acquire a fair knowledge of the fundamental laws underlying the science to which they owe their livelihood" (Paterson, 2003, p. 92).

But this was a long time coming. It was not until after the middle of the eighteenth century that a major conference was held in Scotland, supported by leading industrialists and significantly, the philosopher John Stuart Mill, in which all agreed unanimously that "[t]he time [had] arrived when it [was] desirable and necessary in the education of the people that the principals of science [and technology] should form an important element in the tuition of all classes of the community" (Conference, 1868, p. 17).

A new industrial literacy was deemed essential both for the progression of, and for the participation in, an industrialised society. Craft-based technical education, therefore, began to appear in state funded schooling where, in 1890 in the United Kingdom, it was defined as "instruction which aims at communicating to the pupils [students] knowledge and facilities which have a direct bearing upon some special occupation, industrial or commercial" (Paterson, 2003, p. 90). This focus resulted in the creation of a two-tier educational system predicated upon meritocracy. This directed young people moving from the primary sector into the secondary sector either into a system that was considered academic and that had the development of the individual as its main purpose, or into one which had as its main purpose, a vocationally oriented education in the service of industry. This situation persists in various forms to this day.

In 1762 Rousseau, in his *Social Contract*, foreshadowed arguments against the academic/vocational divide by defining the purpose of education as being between one of two choices. "When instead of raising a man for himself, one wants to raise him for others? [...] Forced to combat [between] nature or the social institutions, one must to choose between making a man or a citizen, for one cannot make both at the same time" (Rousseau, 2002, Book 1, p. 39). More pertinently, however, he goes on later in *Emile or On Education*, to elucidate the view that education should be both comprehensive and non-elitist in nature:

> We must, then, generalize our views and consider in our pupil abstract man, man exposed to all the accidents of human life. If men were born attached to a country's soil, if the same season lasted the whole year, if

> each man were fixed in his fortune in such a way as never to be able to change it – the established practice would be good in certain respects. The child raised for his station, never leaving it, could not be exposed to the disadvantages of another. But given the mobility of human things, given the unsettled and restless spirit of this age which upsets everything in each generation, can one conceive of a method more senseless than raising a child as though he never had to leave his room, as though he were going to be constantly surrounded by his servants? If the unfortunate makes a single step on the earth, if he goes down a single degree, he is lost. This is not teaching him to bear suffering; it is training him to feel it. (Rousseau, 1979, Book 1, p. 42)

Dewey, echoing Rousseau, argued, moreover, that technology education in particular, should be more orientated towards the formation of informed citizens. For Dewey the segregation of vocational education as favoured by industry and commerce struck at the very heart of the democratic process itself.

Dewey's fears were, however, soon to be realised. In 1917 the United States witnessed the passage of the Smith-Hughes Act. This Act promulgated federal support for vocational education that continues up to the present day (Apple & Bean, 1999). Vocational education in the form of technical education in the United Kingdom and industrial arts in the United States thus became embedded in the school curriculum as a subject domain orientated towards vocational education.

## 5 The Problems Encountered with Technology Education's Reformation

It appears that whilst many sincere attempts have been made over the past 40 years to upgrade technology education into a subject that better reflects the technologically textured world we now inhabit, most have not been realised at classroom level. This has come about because curriculum developers, policy makers and politicians have lacked a willingness to make the kinds of changes necessary. They have, rather, maintained a rhetoric that has tended to stick doggedly to orientating technology education within the vocational domain; one involving craft skill development relating predominantly to the perceived needs of industry. By way of an example, I had, in 2007, a chapter published in a book called *Design and Technology for the Next Generation* edited by David Barlex. My chapter was entitled "Vocationalism – Friend or Foe to Design and Technology Education." I have chosen this book in particular because each chapter was subject to a critique by Dr. Peter Toft, the then H.M. government Inspector and Ofsted specialist subject advisor for design and technology in English schools. I have listed below some selected parts of his critique of my chapter:

> [Vocational education] is a part of government policy to improve the skills of students entering the workforce in an increasingly competitive, globalised economy … The aim is to equip them to survive in and contribute to the workforce, through which we create national and personal wealth. One particular facet of this is the current concern in some industrial sectors about the difficulty of recruiting staff to work at craft and technician levels. We will all have heard of recent shortages of crucial specialists such as plumbers. We may also appreciate how these shortages are to an extent being filled with staff from overseas and that some argue this is because we have not trained enough British people to do them …. John Dakers makes a strong case for design and technology to be seen as part of general education and not as a vocational activity …. In Ofsted we routinely observe the subject being taught in this generalised way whilst helping students learn about modern industrial production and its use of resources, together with aspects of consumerism. This links closely with vocational education. In addition, though, design and technology does provide a basis for further specialisation in activities requiring 'know how' as well as 'knowing that' within a range of vocational fields. (Toft, 2007, pp. 277–278)

John Dewey, a much-respected philosopher of education, would have strongly disagreed with Dr. Toft. This is summed up succinctly in the following passage:

> Its [vocational education's] right development will do more to make public education truly democratic than any other agency now under consideration. Its wrong treatment will as surely accentuate all undemocratic tendencies in our present situation, by fostering and strengthening class divisions in school and out … Those who believe the continued existence of what they are pleased to call the "lower classes" or the "labouring classes" would naturally rejoice to have schools in which these "classes" would be segregated. And some employers of labour would doubtless rejoice to have schools, supported by public taxation, supply them with additional food for their mills … [Everyone else] should be united against every proposition, in whatever form advanced, to separate training of employees from training for citizenship, training of intelligence and character

> from training for narrow, industry efficiency. (Dewey in Apple & Beane, 1999, p. 50)

I want to note here that there has been a growing movement towards a modern reformation of a technology education curriculum that is more in keeping with the new and emerging technologies of today, as well as the serious problems facing the globe as the Anthropocene gathers pace. The American Standards for Technological Literacy (ITEA, 2003) for example, advocate the study of the nature of technology, including:
- Technology and its impact upon society.
- Attributes associated with design.
- Abilities that are required to live an informed and meaningful life in a technologically mediated world.
- The study of various modern technologies including medical technologies; agricultural and related biotechnologies; energy and power technologies; information and communication technologies; transportation technologies; manufacturing technologies, and construction technologies.

Publications relating to these aspects include, but are not limited to: *Technology and Literacy in the Twenty-First Century* (Selfe & Burns, 1999); *Technically Speaking: Why All Americans Need to Know More about Technology* (Pearson & Young, 2006); *Advancing Excellence in Technological Literacy* (ITEA, 2003): *Technological Literacy for All* (ITEA, 2006); *Tech Tally: Approaches to Assessing Technological Literacy* (National Academy of Engineering, 2006); *Standards for Technological Literacy* (ITEA, 2007); *Technological Literacy: A multiliteracies Approach for Democracy* (Williams, 2009); *Defining Technological Literacy* (Dakers, 2014a); *New Frontiers in Technological Literacy* (Dakers, 2014b); *Towards a Reconsideration of Technological Literacy* (Hasse & Wallace, 2020).

While all these publications, along with many others, offer a variety of important perspectives on the concept of technological literacy, no universal or unified consensus on the concept appears evident, one that can be articulated into, taught and assessed within technology education curricula around the world. Only partial definitions seem possible. What appears to be viable for the USA does not seem to articulate with technology education curricula in France or Finland or England or Germany. Even local definitions cause dissension.

The USA, for example, has made a supreme effort, one carried out over many years, to identify, quantify, standardise, and, ultimately, assess technological literacy (ITEA, 2003, 2006, 2007). Moreover, they have gone to great lengths to list content standards, student assessment standards and guidelines, together with long range goals and expectations, in order that students might become technologically literate. The three volumes cited above combine to total 436 pages that claim to present "a vision of what students should know and be able to do in order to be technologically literate" (ITEA, 2007, p. VII). Note, however, that the standards aim is to identify what it is that students will know and be able to do in order to *be* technologically literate, suggesting a pre-ordained end point will be reached resulting in technologically literate students. I will present arguments later that will serve to demonstrate that the notion of students becoming technologically literate cannot be accomplished in this manner. Students cannot be taught or told about the impact that technologies may or may not have on different societies or cultures. Students will have to engage with these concepts from their own personal viewpoint and these personal viewpoints, like any set of views, will manifest in a complexity of differing perspectives and world views. Likewise, for similar reasons, they cannot be taught or told declaratively, what specific abilities are required to live an informed and meaningful life in a technologically mediated world. This would pre-suppose a "curriculum-as-plan" as (Aoki, 2005) argues, where the curriculum content is specified in advance. Nor can they be taught or told how to do something procedurally in order to become technologically literate. This again pre-supposes a pedagogy of the transfer of knowledge from teacher (expert) to students (passive recipients). What is needed, instead, is the formation of "a-curriculum-of-lived-experience" (Aoki, 2005) where these issues are discussed and debated openly and democratically; one in which no individual is in charge of the knowledge that is co-produced and subject to constant variation.

The ITEA volumes nevertheless constituted a well-researched set of criteria for the study of technology and for the development of technological literacy in the USA, although they tend to demonstrate a worldview about education that is quantitative in nature. However, these standards have also been subject to change, disagreement, re-evaluation, and discord when it has come to any attempt to implement them within the curriculum. According to Becker et al. (2010):

> currently, 49 of the 50 states have technology literacy goals and standards; more than 80 percent of the states have adopted, adapted, or referenced the International Society for Technology in Education's (ISTE) National Education Technology Standards in state department of education documents. As of 2007, based on a survey conducted by the State Educational Technology Directors Association (SETDA), 21 states reported that they use the ISTE NETS definition (i.e., the six categories of the NETS-S), 15 states reported using a unique state definition, eight states reported using the

> SETDA definition, and seven states reported that they used another method for defining technology literacy. Those varying definitions have been operationalised in the form of technology standards. That is, states encourage the pursuit of proficiency in technological literacy by promulgating student technology standards. There is no shortage of standards for states to adopt or adapt. The International Technology Education Association (ITEA) has developed a series of standards that point out in great detail how one might achieve technological literacy. Those standards, the third iteration of which was released in 2007, include grade-level goals. Additionally, in 2007, the International Society for Technology in Education (ISTE) released the second iteration of the National Educational Technology Standards for Students (NETS-S). (p. 2)

Since these statistics were reported in 2010, the ITEA has changed its emphasis to focus on STEM education, where the Technology and Engineering aspects are seen to work alongside Science and Mathematics. Indeed, in order to emphasise this new STEM orientation, they have recently changed the name of the organisation from the *International Technology Education Association* (ITEA), to the *International Technology and Engineering Educators Association* (ITEEA). The term engineer, however, has become a somewhat ubiquitous term covering many occupations in society today, in that it can now mean craft-based occupations like heating engineer, plumbing engineer, electrical engineer, all of whom are hands-on installation and maintenance occupations.

Professional engineers with university degrees,[2] on the other hand, may also be described as heating, plumbing and electrical engineers, but are more likely to be involved in the design, invention and regulatory aspects associated with their particular field. Whilst these are also considered vocational, they are quite distinct occupations and require very different entry pathways, dependent upon appropriate entry qualifications. As will be seen below, for England, technology education is not considered to be an appropriate qualification for entry to engineering faculties at university level.

Furthermore, the ITEEA have, in 2019, written a rationale for the revision of the *Standards for Technological Literacy* mentioned above. Amongst other things, they acknowledge in this that the concept of technological literacy is difficult to define and is subject to change depending upon cultural conditions. Moreover, citing Zollman (2012), they acknowledge that "there is a difference between literacy and being literate. STEM literacy should not be viewed as a content area but as a shifting, deitic means (composed of skills, abilities, factual knowledge, procedures, concepts, and metacognitive capacities) to gain further learning." They go on, now citing Krupczak et al. (2016) to articulate the concept of technological literacy as a fluid construct that is "constantly evolving or changing" and that, citing Gagel (1997) "to maintain relevance, its content must evolve as a function of changing cultural traditions. The utility of such a literacy would depend on its ability to adapt and keep pace with constant change" (Buelin et al., 2019, p. 5). The report again acknowledges that any universal definition of what constitutes technological literacy is difficult if not impossible to achieve. They conclude, however, that the best way to become technologically literate is to become immersed in technology and engineering contexts as delivered by technology education. On this point, as will become evident later, I disagree. For me, 'understanding about technology,' or the ongoing process of becoming ethnotechnologically literate, must be a priority for all, not just engineers, especially in this technologically textured world we collectively occupy.

These concerns and complications surrounding the delivery of technological literacy are not unique to the USA. In 2017, David Barlex and Torben Steeg authored a publication entitled *Re-Building Design and Technology in the Secondary School Curriculum*, in which they express serious concerns about the status and nature of the subject in England. They argue that "the viability of Design and Technology as a curriculum subject is dependent upon establishing (a) a sound knowledge base for the subject, (b) clarity about its role in the wider curriculum, and (c) an approach to quality teaching and learning" (Barlex & Steeg, 2017, p. 20). Barlex and Steeg are both highly regarded academics in the subject of Design and Technology. The very nature of the title together with the quotation given above does not lend support to the idea of a subject in the best of health. Indeed, overall, the report offers a rather dystopic view of the state of Design and Technology in England. In addition, the report, comprising 49 pages, offers a myriad of suggestions for ways to develop and take the subject forward into the future, which again suggests a subject in need of reform. Significantly, no consideration whatsoever is given to technological literacy as a concept of importance in the re-building of the subject.

Another book authored in 2021 by David Barlex, this time in association with Frank Banks, is entitled *Teaching STEM in the Secondary School: Helping Teachers Meet the Challenge* (2nd ed.). This book explores both the purpose and the pedagogy required for the delivery of STEM education. A very small degree of attention has been given to STEM literacy and a couple of mentions relate in small part to the concept of technological literacy. Neither concept is to be found in the index (Barlex & Banks, 2021).

It appears self-evident then, that the development of technological literacy is not considered an important component of technology education, neither as an individual subject nor as a subject integrated into STEM education. It is also apparent on page 37 of their book which outlines the requirements for teacher professional knowledge (in mathematics, but it can easily be adjusted to technology education), that assessment requirements, content knowledge, subject knowledge, and pedagogic structure are largely, if not entirely, dictated by the imposed national curriculum. The concept of education about technology, made possible by the ongoing process of becoming ethnotechnologically literate, centres, as I have argued, using Vygotsky (1978), around the mediation of technological, socially constructed signs that are external to the individual and which take the form of tools that are both semiotic and artefactual in nature. These come together to form a representation of the technologically textured milieu from which the individual is emerging. Moreover, when these socially constructed signs are individualised, they are subject to change and alteration when the individual engages with another socio-cultural milieu, engendering a multiplicity of interpretations.

As a sign that the craft-based vocational emphasis is currently being given prominence in the United Kingdom, the Design and Technology Association (DATA), is currently promoting what they refer to as the Blueprint 1000 initiative which has been designed to bring education and businesses together. It asks the following question:

> Are you interested in inspiring young people to embark on design, technology engineering and manufacturing careers and to help shape the creative, technical and vocational education they receive? (DATA, 2021)

This initiative explicitly links school design and technology education with industry. It is in the process of developing projects that encourage teachers to pay a fee to spend time in industry as part of their professional development. Several projects are up and running with more to follow. There are short videos on the website which show young apprentices commenting on their workplace experiences. Two videos feature two degree-qualified engineers, both of whom have had several years' experience and are employed by the charity 'Engineers Without Borders UK.' These are the only participants that make any reference at all to technological literacy. Otherwise, the entire Blueprint 1000 initiative is linked to engineering and design vocations, mostly at craft levels. It is a training ground where the emphasis is on training for the perceived needs of industry. This serves as a means to deliver the training needs of industry, as Dewey stated above, supported by public funding in other words, it is the taxpayer that foots the bill.

Giving Evidence to the House of Lords Select Committee on youth unemployment in June 2021, Tony Ryan, the CEO of the Design and Technology

Association (DATA), gave evidence pertaining to the role of technology education in relation to employment. He made a good case for the need for technology education in this respect. However, it is he who is promoting the Blueprint 1000 initiative which is a craft-based vocationally orientated route to employment. In his evidence, he attempted to argue that technology education should be given equal status to academic subjects, hence indicating that the subject is, at present, considered as not being academic. Ryan blamed a new Government initiative for differentiating between technology education, which is considered to be a non-academic, craft-based, vocationally orientated subject and those other subjects that are considered academic, and thus potentially lead towards higher education. This was in reference to the implementation of T-Level qualifications which are described by the UK government as follows:

> T Levels are an alternative to A levels, apprenticeships and other 16 to 19 [year old] courses. Equivalent to 3 A levels, a T Level focuses on vocational skills and can help students into skilled employment, higher study or apprenticeships.
>
> Each T Level includes an in-depth industry placement that lasts at least 45 days. Students get valuable experience in the workplace; employers get early sight of the new talent in their industry. (Government, 2021)

In his evidence Ryan claimed that these T-Levels were unlikely to be accepted for entrance qualifications to universities. This reinforces the message that technology education is not considered an academic subject and is therefore craft-based and vocational in nature. His evidence was as follows:

> Three notable universities to date have made it clear that they will not be willing to view T Levels as suitable entry qualifications to the majority of their degree courses, these being Cambridge, Imperial College and Queen Mary's, London. The Russell Group universities[3] have yet to make a public statement on how they view T Levels which, given the above, I find difficult to see any other way than negatively. As a parent of an able, ambitious young man or woman, why would you encourage your son or daughter towards T Levels given this degree of uncertainty? (House of Lords, 2021)

It is evident that technology education is increasingly becoming promoted as one of the STEM subjects. Science, Mathematics and degree level engineering fall into the 'academic' stream of subjects whereas technology is, as has been discussed, considered to be craft-based and vocational, ostensibly leading towards craft-based engineering careers.

The Engineering stream in STEM education, as delivered by the Technology stream, as Tony Ryan suggests above is, therefore, unlikely to be considered as an entry qualification for a degree-level engineering course such as automotive engineering. It is, on the other hand, more likely to be considered useful for entry into a craft-based vocation such as car mechanics.

The problem with the vocational orientation of craft-based technology education is that it claims to promote technological literacy as an important component, but in reality, avoids incorporating it into the taught curriculum. Rasinen, for example, undertook a systematic analysis of the curriculum content for technology education of six countries and found that "although the format and approach in the six curricula studied differ from one another in many ways, common features were found" (Rasinen, 2003, p. 42). These common features found were, however, situated within technology education programmes that were at very different stages of development, resulting in the conclusion that, "a single model cannot be applied to each country" (Rasinen, 2003, p. 45). Significantly, although Rasinen found that the concept of "technological literacy was a universal goal" for all curricula (Rasinen, 2003, p. 45), it is a goal, I argue, that has never been fully realised.

In contrast to the conclusion offered by Rasinen, is the concept of an international technology education baccalaureate as proposed by Williams (2008). At present this is a functioning international program for the delivery of a universal and unified form of technology education around the world, one that bears a distinct resemblance to Bacon's version of what could be described as a universalised technology education paradigm. Any school choosing to adopt this program does so on a voluntary basis. Williams argues that "[t]he reasons for schools adopting the IB [International Baccalaureate] vary, but the main reason is the provision of an internationally accepted pre-university certification that is reputable and transferable" (Williams, 2008, p. 53). However, the International Baccalaureate Organisation [IBO] syllabus for the subject Design Technology, as it is known in this context, does not appear to feature the concept of technological literacy as an area of study, at least not in the subject areas listed. "The aim of the [...] design technology course is to foster the skill development in students required to use new and existing technologies to create new products, services and systems" (IBO, 2021). Clearly there is an emphasis on design and indeed design literacy which claims to have an ethical framework. However, design in technology education is an applied skill; it is something that humans do, for the most part to create artifacts intended for consumption. This implies designing for profit in order to create a buoyant economy. This economic aspect of design does not exist in school-based design, Baccalaureate or otherwise. This renders the ethical element somewhat redundant. To work as a designer of technological products successfully, whether self-employed or as part of an established organisation is done in order to earn a living. The ethics of designing products that do not contribute to a global crisis of climate change are thus in conflict with the creation of economic stability or buoyancy. The pedagogy that I am proposing is designed to address issues such as this. There is no evidence of this type of pedagogy being applied in extant models of craft-based technology education, nor in the curriculum of the IBO for design and technology education. The pedagogies currently employed in school craft-based technology education do not allow for any truly critical analysis to take place because they are principally assessed on the acquisition of *a priori*[4] knowledge and skill development. Ethical issues constitute a moving feast that differs over time, cultures and worldviews, and so can be assessed only in terms of *a posteriori* knowledge and skill development. This is where IBO assessment is flawed. It claims to implement "a rigorous assessment process in order to ensure standards remain high." In order to achieve this stated goal "the assessment criteria are standardised [in order that] the aims and objectives are achieved." It is principally in this respect that I part company with the concept of a universal technology education paradigm, especially one that may purport to include the concept of technological literacy. Standardised assessment protocols, by their very nature, serve to problematise pedagogies that involve critical thinking development, a pedagogy, that is, in my view, essential for the development of becoming ethnotechnologically literate. Critical thinking about technology involves the development of perspectives that are immanent, multiple, value-laden, and temporal with respect to technological/social/cultural relationships, whether existing or potential. These relationships are formed within the parameters of socially constructed milieus that are complex and local, something that universalised and standardised assessment procedures find difficult, if not impossible, to measure. Standardised assessment protocols, especially, although not limited to, those that involve answers to multiple choice questions, require overwhelmingly conformist criterion within which local interpretation has no place. Only "this or that" responses predicated almost exclusively upon the development of procedural and declarative knowledge prevail.

These issues have accounted for long-standing problems in the delivery of a more relevant and up-to-date education about technology. The subject continues to be frozen in time as the following citations confirm:

> in order to address the need for teaching technology for technological literacy in contemporary society, technology education must be more than the teaching and learning of handicraft skills, as was, and often still is, the

> case in classroom practice. Such education requires that pupils acquire an understanding of what technology is and how technological developments take place. It also means that a more theoretical component must be added to the practical dimension that already exists. In other words, not only skills, but also concepts of technology need to be taught and learnt. (De Vries & Tamir, 1997, p. 3)

> School education seems to be mostly stuck in an outdated industrial era [worldview], unable to sufficiently address the significance and increasing rapidity of changes to humanity that are upon us. An integrated forward-looking view should, now more than ever, be of central importance in how we educate. (Gidley & Hampson, 2005, p. 255)

> Within [what is a] behaviourist inspired metaphor, the learner is a passive being whose repertoire of behaviours is determined by rewards and punishments encountered in the environment. The metaphor … has straightforward implications for instruction [in technology education], namely, creating situations that elicit responses from learners and providing appropriate reinforcement for each response. Drill and practice is the epitome of instruction within this view of learning. (Mayor, 1992, p. 407)

> [Technology] education is commonly perceived as exclusively formal, confined to traditional classrooms for children or adults alike, and focussing on rigid measurable objectives, even if under several different guises depending on times and political contexts. The fashionable term lifelong education is reduced to professional training and acquiring technical skills to feed back into contemporary knowledge economy …. [It is] the neglect of real-life topics that would have called forth critical and reflective thinking [that] pervades the present system of education. (Semetsky, 2015, p. 1069)

These arguments are not just limited to technology education in school, as Goodman elucidates:

> Technology must have its proper place on the faculty as a learned profession important in modern society, along with medicine, law, the humanities, and natural philosophy, learning from them and having something to teach them. As an [ethical] philosopher, a technician should be able to criticise the programs given him to implement. As a professional in a community of learned professionals, a technologist must have a different

> kind of training and develop a different character from what we see at present among technicians and engineers. He should know something of the social sciences, law, the fine arts, and medicine, as well as relevant natural sciences. (Goodman, 2011, pp. 40–41)

A two-year longitudinal European research study, looking into ways to increase participation in STEM subjects across Europe, particularly by women, was carried out with the European Commission between 2001 and 2003. One consensus that was reached in the concluding report, was that the delivery of science and particularly technology subjects at school level, were in much need of major reform in terms of their delivery. This was reported in a paper written by Dow (2006) entitled: "The Need to Change Pedagogies in Science and Technology Subjects: A European Perspective." Dow reported, as the title of the paper suggests, that it was essential that new forms of pedagogy be implemented across Europe, in both science and technology subjects at school level. Dow (2006) goes on to outline the problem of the extant pedagogical models involved in the delivery of craft-based technology education across Europe, all as concluded in the study.

> The mechanistic processes which underlie the dominant existing model have the effect of reducing technological knowledge to small discrete components which are learned, mostly through drill and practice, and subsequently tested in situations completely devoid of any meaningful context. This was supported from the research which indicated that technology education across Europe had a vocational emphasis which tended, for the most part, to be directed towards skill acquisition in the service of trades occupations, and was taught in whole class settings with children carrying out essentially the same tasks. This type of instruction in turn results in a depersonalisation and fragmentation of the child's experience, and in its "whole class," assembly line production of skills, estranges children from not only the subject and each other, but from themselves as well. (p. 311)

Considered from the perspective of the many who have the power to design, regulate, and implement the delivery of craft-based technology education, it becomes apparent from the way it continues to be delivered, that the social construction of craft-based technology education, as recognised by students, parents and particularly females, is that it is an initial training ground designed to meet the perceived needs of industry, an industry that is gender biased towards males. This clearly has particular implications for the recruitment of females into technology education. Murphy, moreover, concluded that this is as a direct result of school-based knowledge construction:

> [I]n the discourses that circulate about technologies, women's and girls' technical competence is denied and beliefs about their 'lack' and men's and boys' 'expertise' become part of our 'common-sense' knowledge about the way people are. Women's involvement in technology continues to be problematic with product design, decision-making, and development dominated by men. (Murphy, 2006, p. 219)

The difference between the sex of the male and the female, purely from a physical basis is relatively obvious and biologically determined.[5] The difference from a cultural gender standpoint, however, is fraught with difficulties. The cultural norms for gender vary across communities and countries. These norms also vary across time. "The roles of males and females are almost always socially constructed and not biologically determined" (Slattery, 2006, p. 156). Paetcher (2007) highlights the fact that craft-based technology education seen as a craft-based vocational subject, is conceived very much as a masculine working class domain. This, she argues, leads

> young women overwhelmingly [to] opt for pathways that lead to poorly paid jobs with little possibility for advancement or the development of their skills, in contrast to young men, who tend to favour these courses leading to recognised trades qualifications with the potential for self-employment and good rates of pay. (p. 129)

Murphy (2007) reports similar findings. She highlights the fact that pupil choices in the technology subjects follows gender stereotypes: a significant majority of boys elect to take the workshop based or electronically orientated subjects whereas a significant majority of girls elect to take food or textile orientated subjects. This is in exact contrast to the performance indicators in examination results. Statistically and paradoxically, more girls achieve higher grades in examination results in virtually all technology subject domains. It is not, therefore, a question of ability. Murphy observes that that these gendered roles in craft-based technology education offer us clues as to pedagogies that might be better employed in the delivery of the subject.

The questions raised above may not only shed some light on the situation but may help to deconstruct it somewhat. Being perceived as a subject that is more orientated toward serving the needs of industry clearly locates craft-based technology education as an initiating base for training future workforces.

This, in turn, has far-reaching implications for the development of critical capacities:

> If knowledge development in technology education is perceived by students to be purely neutral and functional, devoid of any ethical, political or hermeneutic qualities, they are more likely to develop into adults who will hold an uncritical and restricted view of the technologically mediated world they occupy. (Dakers & De Vries, 2009, p. 206)

Becoming technologically literate, or multiliterate as Williams argues, requires a more complex and progressive learning space: one that differentiates between the concepts of a universal technology education Baccalaureate as offered by the IBO, for example, and that of the very much less instrumental and much more democratic educational concept of technological multiliteracy. Williams suggests that

> if technological multiliteracy becomes the focus of technology education through its positioning as a moral rather than vocational or instrumental practice, and the mechanism is available for students to express their beliefs, then the move toward a more democratic technological order becomes possible. (Williams, 2009, p. 252)

Whilst I lean strongly toward this perspective, I take issue with positioning multiliteracy exclusively as a moral issue. By way of clarification, I notice that the terms moral and ethical often tend to become conflated, and while I make no claims on Williams' use of the term, I consider the distinction important. I argue that morality is more orientated towards some set of imposed hierarchical regulation, 'though shalt not ….' We hear often that 'they are immoral,' implying that 'they' are bad people who do not follow rules or conventions. The hippies in the sixties were often categorised thus. But we rarely hear that 'they are unethical.' Unethical is usually associated with an action and less so with a person or people. 'I think what you are proposing is unethical' (see Jun, 2011, p. 124, for example).

## 6 A New Age for Ethnotechnological Literacy

It becomes clear then, that a genealogy for craft-based technology education reveals it as rooted firmly within an industrial, craft-based vocationally-orientated past where its purpose was clearly established. It served as an introduction to initial skill procurement for boys entering industry, and to housekeeping skills, in the form of domestic science or home economics, for girls. This emerges out of the post WW2 era. Although over the past few decades, a new rationale for technology education has been promoted, it is also clear that, in many cases, the older industrial model persists. This is not the fault of teachers or schools. Governments and capitalist institutions make grand declarations about their commitment to technological literacy as part of the technology education paradigm, then continue to ignore any progressive research into how that might be achieved. This obfuscation continues whilst institutions that design, invent and create complex technologies today, (and there are many), create technologies that have an ever-more palpable political, social and cultural impact upon individuals and societies around the world. Whilst these institutions can be considered, on the surface, to be digital technology producing institutions, even the outcomes of knowledge-producing institutions, if not well regulated, can be subject to the type of external abuse that impacts upon all levels of society. Surreptitiously woven into the folds of new, 'must have' technologies, lie so many unforeseen potentialities that are open for integration into societies, virus like, such as hacking, tweeting, child paedophile groups, election manipulation frauds and surveillance technologies to mention but a few. This abuse is most certainly not done for the good of humanity, but to enable what Deleuze calls 'societies of control.'

"[S]ocieties of control operate with machines whose passive danger is jamming and whose active one is piracy and the introduction of viruses" (Deleuze, 1995, p. 180). Craft-based technology education in schools, however, has little exposure to modern technological development in actuality, politically or philosophically. Craft-based technology education continues to teach young people about simple skill sets: levers, pulleys, woodworking, metalworking, graphics, cooking all under the guise of 'design and technology.' In the UK the curriculum guidelines claim this subject area is the ideal platform for pupils who would like to enter the construction field in careers such as Architecture, Civil Engineering, or other types of engineering including Mechanical Engineering and Product Design. But universities, as has been previously noted, do not accept craft-based technology qualifications as an entry passport to these subject areas. Past papers for all aspects of design and technology education offer complete past papers from a web site called revision world,[6] where students can revise answers to questions that are commonly asked in an examination. Inspection of these reveals that virtually all questions are linked to either procedural or declarative knowledge domains: domains in which pre-existing knowledge transferred from teachers to students who are subsequently expected to reproduce it in the form of summative examination, as in Freire's banking method. I was unable to find any questions relating to sustainability, global warming, technological impact or any of the issues that I have so far discussed. This suggests a society of control which, in a school system propagates: "continuous forms of control [resulting also from curriculum developers], and the effect on the school of perpetual training, [together with] the corresponding abandonment of all university research, [with instead, the] introduction of the 'corporation' at all levels of schooling" (Deleuze, 1995). Deleuze sums up societies of control thus:

> The conception of a control mechanism, giving the position of any element within an open environment at any given instant (whether animal in a reserve or human in a corporation, as with an electronic collar), is not necessarily one of science fiction. Felix Guattari has imagined a city where one would be able to leave one's apartment, one's street, one's neighbourhood, thanks to one's [...] electronic card that raises a given barrier; but the card could just as easily be rejected on a given day or between certain hours; what counts is not the barrier but the computer that tracks each person's position – licit or illicit – and effects a universal modulation. (p. 182)

The process of becoming ethnotechnologically literate requires participants to become sensitive to the impact of new and emerging technologies upon their technologically mediated lives. Moreover, it requires them to engage in a more critical examination of the interplay between technology, the environment and society.

The crucial importance of this can be highlighted by consideration of a recent report entitled *The Economics of Biodiversity: The Dasgupta Review*. Published in February 2021 this had a preface written by David Attenborough, some of which is reproduced below:

> We are facing a global crisis. We are totally dependent upon the natural world. It supplies us with every oxygen-laden breath we take and every mouthful of food we eat. But we are currently damaging it so profoundly that many of its natural systems are now on the verge of breakdown.
>
> We are now so mechanically ingenious that we are able to destroy a rainforest, the most species-rich ecosystem that has ever existed, and replace it with plantations of a single species in order to feed burgeoning human populations on the other side of the world. No single species in the whole history of life has ever been so successful or so dominant.
>
> But consider the following facts. Today, we ourselves, together with the livestock we rear for food, constitute 96% of the mass of all mammals on the planet. Only 4% is everything else – from elephants to badgers,

> from moose to monkeys. And 70% of all birds alive at this moment are poultry – mostly chickens for us to eat. We are destroying biodiversity, the very characteristic that until recently enabled the natural world to flourish so abundantly. If we continue this damage, whole ecosystems will collapse. That is now a real risk. (Selected paragraphs from the preface by David Attenborough to Dasgupta, 2021, pp. 1–2)

The Dasgupta review reveals the many problems that are in place resulting from technological development, especially in the epoch of the Anthropocene.

> The incentives entrepreneurs have for developing technology are shaped by the systems of property rights in place. Remarkable post-war developments in sonar technology and advances in the technology for harvesting fish came about in large measure because ocean fisheries beyond national jurisdiction are free. Unbridled application of modern technology in clearing tropical rainforests has been made possible because governments have permitted it at a small price. Both forms of environmental destruction would have been avoided had institutions not failed. There is nothing good or bad about technology per se, it is the use to which it is put that affects [the world] .... But no matter how effectively institutions are established to synergise with technological advances, unending growth in global output is an ecological impossibility. The biosphere is bounded and there are theoretical bounds on global output. (Dasgupta, 2021, p. 124)

The report highlights issues in modern technological development, some of which are incentivised by way of Government sponsorship, such as the installation of solar panels, thereby reducing carbon emissions. What is not made clear, however, is that solar panels are built using aluminium, zinc, cadmium, and other minerals. And mining and quarrying usually require that forests be destroyed (Dasgupta, 2021, p. 124). Figure 1 rather elegantly demonstrates Dasgupta's recommendations for what needs to happen between Humanity's demands and Nature's supply.

Notice that Dasgupta recommends a need for an educational reformation; one that has at its core the creation of an informed citizenry; one that is empowered to make informed choices about the way the human/technology relationship is able to resolve the extant exploitative affiliation it has with the rest of the natural world, to one that is both viable and sustainable.

In addition, a new book entitled *Breaking Boundaries: The Science of Our Planet*, published in 2021 and subsequently made into a Netflix documentary

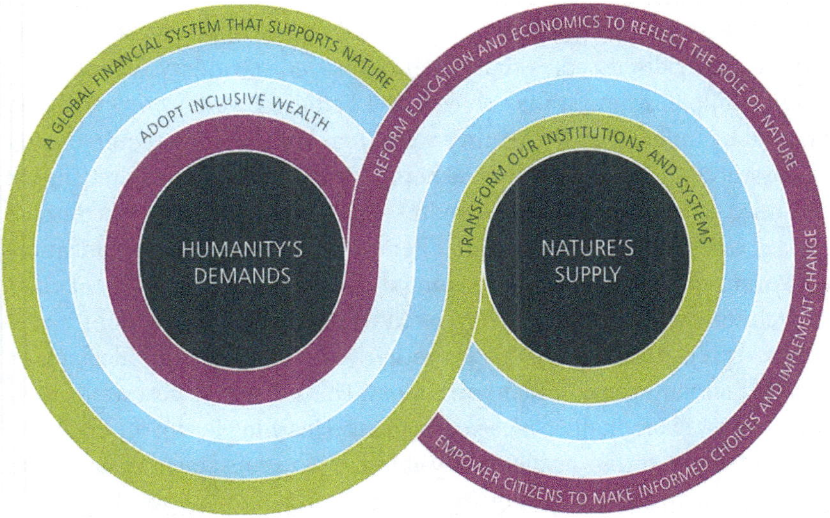


FIGURE 1 Creating an enabling environment (from Dasgupta, 2021, p. 493)

highlights, like the Dasgupta report, the fact that we are on the brink of a critical moment in human history. This will, it suggests, require a rethinking of our relationship with our planet. It consequentially tries to offer a new course for our future. It also provides a disturbing look at what is happening to our planet (Rockström & Gaffney, 2021).

If these recent reports and books do not, in themselves, highlight the way that human beings together with their technologies are destroying the planet, one only has to be aware that in (2021) a series of devasting events happened with forest fires and flooding in many different locations around the world. Education about technology simply cannot continue to turn a blind eye to this. A complete rethink of the current pedagogical framework is urgent and that will be the theme that the following chapters on education about technology will pursue.

## 7 Concluding Remarks on This Chapter

It is evident then that the development of (ethno)technological literacy is an area that is profoundly lacking in the delivery of craft-based technology education today. The craft-orientated emphasis seems to persist dogmatically. This can be attributed to several reasons, many of which I have outlined above. Another is the general public perception of what craft-based technology education actually is. It is a practical subject that a significant number of the today's adult population studied when they themselves attended school. It is therefore a subject that is not perceived to have changed much. Another reason again, one that is rarely discussed, is that when I attended secondary school in the 1960s for example, amongst the subjects that I studied, were woodwork, metalwork and technical drawing, as they were then known. Each required dedicated classrooms, in the form of workshops for the first two and a drawing lab for the third. These dedicated classrooms, unique to each subject, continue to this day where schools continue to offer craft-based technology education. Rote learning continues to be the dominant pedagogy in the delivery of craft-based technology education, a pedagogy in which the development of '*knowing how to*' (procedural knowledge), together with '*knowing that*' (declarative knowledge) remains the dominant orthodoxy. This is because post war technical education was originally conceived as being a learning forum for boys who were considered to be less academically able and who consequently needed to be educated for trades-based industry. A technical education would prepare them for their inevitable transition from school to industry. Technical education was thus industry based, offering a pre-apprenticeship form of industrial education that was trades-based. This resulted in the recruitment of teachers who came from trades-based occupations. Although technology teachers with degrees are now required by the profession in many countries, the emphasis in secondary school education is still orientated towards workshop practice. Given then, that craft-based technology education continues to have an emphasis on workshops in woodworking, metalworking and plastics, it cannot be an environment to let young people simply enter and experiment. They must learn to use tools and machines safely. This means that the only pedagogies that can be employed are those of a declarative and procedural nature: in other words, transmission of knowledge from an expert (the teacher) to the novice. It is only when design enters the equation that experimentation and creativity can be explored. The creative element is, however, almost always secondary to the skills element and is, again for the most part, a prescribed route for producing a standardised folio schematic. The trouble with the folio, that is the design component, is that it, too, is largely prescribed when it becomes formally assessed. Folios have to comply with a given format, the outcome of which will then be the fabrication of the given design. If a student were to design their project on the back of a cigarette packet, so to speak, and go on to produce an outstanding product, they would, nevertheless, lose marks for their design folio. A cigarette packet does not follow the prescribed curriculum protocol.

An interesting variation to this model was offered by Barlex who, in his Nuffield "Young Foresight Project" removed the requirements of fabrication, thereby enabling students to "concentrate on designing without being limited by their personal making skills and the facilities available in the school" (Barlex, 2014, p. 126). He went on to argue that this progressive pedagogy also encouraged students to act collaboratively by discussing their ideas with each other. In other words, the assessment protocols adopted did not compel them to act independently of each other. Alas however, these innovative practices were never fully taken up by schools, and it also seems that this project has been abandoned. This represents yet another example of policy makers ignoring or rejecting research led initiatives for technology education.

It is for the reasons outlined in this chapter that I advocate a paradigm shift in education about technology, to one that utilises new forms of critical pedagogy and is delivered as a separate subject from the craft-based technology education subject that continues to this day. This would serve to disentangle two discordant and incompatible pedagogical methodologies from each other, thereby enabling each to go their own way unencumbered by the other.

By attempting to graft the concept of technological literacy on to craft-based technology education, is trying to make it do more than it is capable of. It would, in my opinion, be better to teach what David Pye (1968, 1978) refers to as *The Nature and Art of Workmanship*, and *The Nature and Aesthetics of Design*. This involves learning to design and manipulate materials with tools and machines and thus supports being taught by way of declarative and procedural knowledge pedagogies.

The social aspects of the concept of ethnotechnological literacy or the development of critique relating to the human technology relationship cannot be considered as an add-on, as is the case for the most part, if and when it is applied. An appreciation of the relationship between human and technology has always been a vital component of the survival of the human race, socially, culturally, psychologically and physically. Without technology it is not likely that the human race could have evolved and survived. Ironically, there is a distinct possibility that the development of technology, particularly those related to weapons of mass destruction and global warming, may well also, ultimately, result in the extinction of the human race. As I write this chapter, the temperature outside is minus ten degrees. I am very happy to have the technologies of a warm fire, a well-insulated house and a woolly jumper to offer additional comfort. I have, however, also been reading in the press about the American election and the way that technology has been used to manipulate Trump supporters into thinking the election results were rigged, a perspective that led to violent outcomes. These methods of persuasion are made possible by digital technologies. It simply beggars belief that, in this globalised world, there does not appear to be an established educational forum to address the human/technology relationship, a relationship that clearly, not only has a major impact on human life, but which forms an integral and necessary component for the ongoing survival of the human race. Curriculum developers take note.

Stiegler suggests that technics should most properly be thought of as the key philosophical question of our time. Perhaps we should expand this to include education about technology as being the key philosophical/educational question of our time.

Given that there are social, cultural and political dimensions inherent within educational policy-making, technology education, particularly craft-based technology education is in urgent need of reform. The curriculum requirements as mandated by successive governments around the world, continue to be responsible for causing a two-tier system of education, one predicated upon the academic/vocational divide and technology education is most certainly considered to reside on the vocational side. This is made manifest in the latest governmental reforms that are now promoting the concept of STEM education as discussed above, where Science and Mathematics are both considered academic subjects whilst Technology remains a vocational subject orientated towards industry. It is also significant that Science and Mathematics are considered to be two of the essential entry qualifications for most university Engineering departments. As long as the perception of technology education remains rooted in this craft-based workshop paradigm, predicated upon rote learning and transmission models of teaching, which it has done since its inception, it will suppress any truly creative and higher-level thinking development.

It is for the reasons given above that I argue, that for learning *about* technology in the 21st century to be creative and innovative, the ongoing process of becoming ethnotechnologically literate needs to have a new model of pedagogy; one that involves the active participation of students "thinking through technology" (Mitcham, 1994). To that end I will, in the following chapter, discuss the philosophical foundations for a 'nomadic pedagogy' one that is much influenced by Gilles Deleuze, Felix Guattari and Gilbert Simondon (Deleuze & Guattari, 2004; Simondon, 2017). This will take the form of a new paradigm for education about technology. Ethnotechnological literacy involves recognising that our perceptions of technology involve a multiplicity of concepts, virtually all of which are subject to a multiplicity of competing forces. Simondon refers to these multiplicities as modes of existence. Modes of existence shape the way we perceive the world. A critical pedagogy in schools, as Giroux sees it, "are modes of intervention dedicated to creating those democratic public spheres where individuals can think critically,[7] relate sympathetically to the problems of others, and intervene in the world in order to address major social problems" (Giroux, 2020, p. 12). This critical pedagogy enables individuals to observe the world through the lens of other individual perspectives, thereby opening up a richer variety of ways to perceive the world. Transmission models of education, in contrast, are "limited to the propagation of a culture of conformity and the passive absorption of knowledge" (Giroux, 2020, p. 3). What becomes evident then, is that craft-based technology education and education about technology, or ethnotechnological literacy, require two structurally different forms of pedagogy.

## Notes

1 Not dissimilar vocationally orientated institutions were evident in other Occidental countries.

2 I make this point to demonstrate that the term 'engineer' has many designations today and does not signify one specific occupation but many. I am not suggesting that one occupation is in any way superior to another. Professional engineers and craft-based engineers are all vocational occupations.

3 The Russel Group universities in the UK are considered to be elite universities, much along the same lines as the Ivy League universities in the USA.

4 A priori knowledge is knowledge that is gained independent of experience. A posteriori knowledge, on the other hand, is knowledge that is dependent upon experience. Well known examples are that 'all bachelors are not married' does not require that the knower has experienced marriage to know this but will come to this knowledge from other sources. Whereas, to know 'it is raining outside' is to have experience of the weather.

5 I accept that LBGTQ communities form a diverse set of populations who do not necessarily fit within a simple binary distinction. However, I only wish to make the point that human beings, all human beings, have not significantly changed physically over time, as much as they have changes socially and culturally. There is a fundamental difference between the concept of sex and the concept of gender.

6 See https://revisionworld.com/gcse-revision/design-technology/gcse-design-and-technology-past-papers-0/ocr-gcse-design-and-technology-past-papers

7 Critically in this sense is not about finding fault, *per se*, but being able to reflect meaningfully through careful analysis and judgement. It is in this sense that, as Paulo Freire in his classic book *Pedagogy of the Oppressed* argues, that education can never be neutral, it is either an instrument of liberation or an instrument of domestication.

CHAPTER 4

# What Are the Philosophical Foundations for a New Nomadic Pedagogy?

One never becomes ethnotechnologically literate; one is constantly in an ongoing process of becoming ethnotechnologically literate, just as one is constantly in a state of becoming something other (baby becomes child, becomes adolescent, becomes adult, becomes parent, becomes grandparent, etc.). This applies as much to technologies as it does to human beings. In this chapter I will outline the philosophy of 'becoming other' as Deleuze and Guattari, and Simondon propose. This concept sits at the heart of the concept of 'nomadic pedagogy.' However, prior to that I want to discuss my reasons for using the philosophies of Gilles Deleuze and Gilbert Simondon as my main influences for this thesis.

I have been studying Deleuze for some time now and have no hesitation in stating that he is not an easy philosopher to comprehend. For this reason, I have offered my interpretation of Deleuze in a way which I hope will make his thinking more accessible to those who are not acquainted with his philosophy. I have referred to him in several of my previous publications regarding technology education, for two main reasons. First, I have always had an interest in philosophy; an interest that became more specialised when I started to work in the university sector, teaching new teachers about technology education. My interest in the philosophy of technology grew, and as I conducted more and more research into technological literacy in particular, I found that I could conflate the three concepts that most interested me: philosophy of technology, technological literacy and the philosophy of Gilles Deleuze. My second reason is that the philosophy of Deleuze, though highly relevant, very rarely appears in research related to technology education. Indeed, the highly regarded Deleuzean philosopher of Art Education from Canada, Jan Jagodzinsky, states that in terms of noticing any references to Deleuze in technology education research, "The exception is perhaps John R. Dakers and Stephen Petrina [who are both] attempting to introduce Deleuze and Guattari into the mainstream technological education, but it all seems that it is in its beginning stages" (Jagodzinski, 2017b, pp. 25–26). It is for these reasons that I am introducing Deleuze's thinking to my colleagues in technology education, and particlularly research into technology education. As will been seen in this chapter and beyond, Deleuze enables the introduction of a completely new philosophical discourse into the field of technology education; one for which I, and many

 | DOI:10.1163/9789004537002_004

others involved in the philosophy of education, believe whose time has come (see, for example, Jagodzinski, 2017b; Petrina, 2020; Protevi, 2001; Sauvagnarges, 2016; Semetski, 2006). I have also included of a number of philosophers who either had an important influence on Deleuze (Simondon and Nietzsche) or who have, in turn, been influenced by Deleuze or Simondon (Stiegler) or have produced concepts very similar (Van Riessen). The following brief reviews on Simondon and Stiegler form part of my two chapters on each philosopher in Dakers et al. (2019). The reviews of Nietzsche and van Riessen are original to this book.

## 1 Gilbert Simondon

Gilbert Simondon was a French Philosopher of technology from the post war period up until his death in 1989. Only two principal works were, however, published while he was alive: *On the Mode of Existence of Technical Objects* [*Du mode d'existence des objets techniques*] and *L'Individuation à la lumiére des notions de forme et d'information* [*Individuation in Light of Notions of Form and Information*].

Simondon was motivated by a desire to raise awareness of the meaning of technical objects at a much deeper level than the study of industry and the mechanical arts. He also offers his version of the genesis of technicity which has a bearing on craft-based technology education. Simondon consider the genesis of human being as taking place in an originary phase. It is not important to know exactly how long this phase lasted or when it began: it is important only in that it represents the phase-shift from pre-human existence to the originary phase that accommodated the genesis of human being. Simondon uses this point in human evolution as his starting point to develop his philosophical thesis *On the Mode of Existence of Technical Objects*. The original mode of existence, in terms of the concept of the human being, emerges from what he refers to as the magical mode. The magical mode, for Simondon, represents the original phase from which human being emerges. It originates as a primitive mode, where the relationship between the human and the rest of the world is indivisible. This original phase exists as a unified milieu, a world in harmony in which there is no distinction made between subject (human) and object (everything else). There are only natural causes. The evolution of human being for Simondon, in this phase, starts out as purely biological. It is a unique mode of existence which forms not only the genesis of human being but the genesis of technicity. However, within this original phase, transitions occur that, over time, eventually mark the transformation from a purely biological evolution to a psychic and collective one. It is in this sense that the magical mode acts

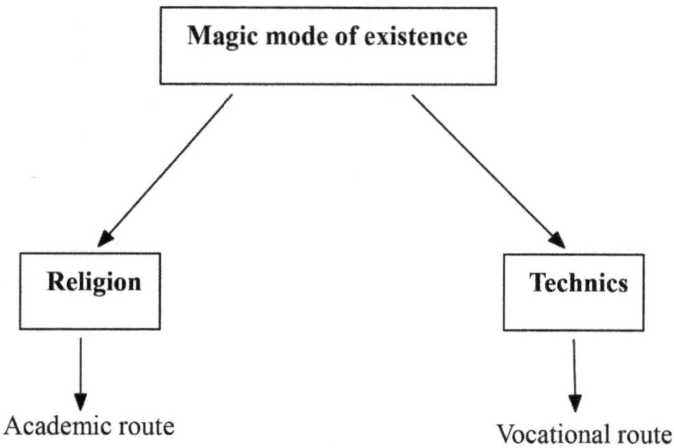


FIGURE 2 Separation of culture and technics

as a threshold between nature and culture which results in a differentiation between object and subject, between thought and action (Bardan, 2015).

The evolutionary process of any individual is what Simondon refers to as the process of individuation. It is important to note, however, that individuation refers to the processes that explain the coming into being of everything, whether material, organic or human and whether regarded as subject or object. It does this by "providing models for understanding how things, including living beings, are brought into existence as cohesive individuals. Simondon opens up new ways of understanding identity, transformation and creation – all central ingredients in a radical reconceptualisation of thought" (Grosz, 2013, p. 37). Thus, for Simondon, an original individuation, or phase shift takes place, from what he refers to as an original magic mode of existence. This magic mode where all living things lived in harmony with the natural world, develops over time, as some humans begin to develop artificial lifeworlds and others begin to question the concept of existence. This eventually separates into two new phases involving a differentiation between nature and culture, subject and object, thought and action, resulting ultimately in a radical break in the unity of the magical mode into two separate and to some extent, opposing modes of existence. The cultural/subject/thought mode became religion/philosophy, whilst the nature/object/action mode was made manifest in technics.

> We suppose that technicity results from a phase shift of a unique, central, and original mode of being in the world: the magical mode; the phase that balances out technicity is the religious mode of being. Aesthetic thought appears at the neutral point, between technics and religion, at

> the moment of the splitting of the primitive magical unity: it is not a phase, but rather a permanent reminder of the rupture of unity of the magical mode of being, as well as a reminder of the search for its future unity. (Simondon, 2017, p. 174)

This dual phase shift from the unity of the magic mode of existence epitomises the genesis of what, in educational terms, is now referred to as the academic/vocational divide. But what constituted this splitting away from the unified magical phase in the first place? For Simondon, the magical phase acted as a vital liaison between human being and world, their primitive unity being made without distinction of subject and object. The universe in this sense is experienced as an associated milieu. Through time, the magical phase is composed of many phase shifts resulting from the very many relationships encountered between human beings and their particular environment. The physical activities of experiencing mountains, summits, gorges, the heart of the forest, hunting, eating, simple tool-use, sit alongside the metaphysical concepts of inaugurations, community, dogma etc., all of which conspire to form separate structural systems.

> The key-points of structure separate and objectivise themselves; technics turns it into [physical activity] and technical objects become functional, instrumental, whereas the powers of the [metaphysical] are subjectivated under the form of the divine and the sacred (gods, heroes, priests). A distance is introduced between human being and world. This distance is mediatised by technics on the one hand, and religion on the other. Where there was only a unity of the living being and its milieu, a difference between man and world appears. (Michaud, 2013, pp. 122–123)

Ultimately, these two phases become so polarised that they can no longer survive together as part of a magical unity. The consequence is that they split away from the unified single mode by forming two new separate modes: religious subjectivity and technological objectivity. These twin tracks mutate further by incorporating social and political thought into the religious sector and evermore complex technics into the human world. The religious, social and political become metaphysical in essence, more associated with abstract thinking: ergo the mind. These essential qualities are considered to be transcendental, dogmatic, idealistic and hierarchical. The technical, on the other hand, is made manifest in its actuality and is more associated with physical activity: ergo the body. It becomes a technological prosthesis in the service of human being made possible by exploiting the material world. The religious, social and political guide humanity. The technical serves humanity and so, by logic, does nature. It is from nature that the physical aspects of technology emerge. Whereas unity belongs to the magical world, "[t]he phase shift opposing technics and religion irreducibly leaves the content of technics with a status lower than unity and that of religion higher than unity" (Simondon, 2015, p. 185).

Bernard Stiegler, moreover, reinforces the longevity of this lack of status:

> At the beginning of its history philosophy separates technē from ēpistēmē ... It is the inheritance of this conflict – in which the philosophical ēpistēmē is pitched against the sophistic technē, whereby all technical knowledge is devalued – that the essence of technical entities in general is conceived. (Stiegler, 1998, p. 1)

This distinction continues to this day, in that the status of craft-based technology education, which is clearly embedded in the technics mode of existence, is considered inferior to the status of academic subjects, which are clearly embedded in the religious, social and political mode of existence. This distinction, moreover, continues to resonate as strongly as it did, as depicted by Simondon's genesis of human being.

It is in this sense that I make a distinction between craft-based technology education as it is taught today, a subject that stubbornly continues to sit within the technics mode of existence and what I refer to as education *about* technology or ethnotechnological literacy (Dakers, 2006, 2007a, 2007b, 2014, 2018).

## 2 Friedrich Nietzsche

Another philosopher that had a strong influence on Deleuze was Nietzsche. He was born in Röcken a small village in Saxony, in what was then Prussia. His father was a Lutheran pastor who died when Nietzsche was only five years old. He wrote many books and other publications and was a great admirer of the philosopher Arthur Schopenhauer, who became a great influence. Without doubt, Nietzsche was a great philosopher, albeit a controversial one. He constantly challenged the received wisdom of the day. He is probably best remembered for his pronouncement that "God is dead." However, for Nietzsche, this was not an argument that atheism was true. It is perhaps understood better when the whole pronouncement is heard: 'God is dead, and we killed him.' Nietzsche was living during the Enlightenment, a time when Europe was going through an incredible transformation. Whereas before, the way of the world was explained by the church, the rise of science resulted in moral scepticism which had the effect of essentially undermining Christianity, promoting instead a secular society; hence God is dead, and we killed him. Nietzsche's corpus of work is huge and diverse and his relevance to this book is due to his strong influence on Deleuze. Deleuze was inspired by Nietzsche particularly with respect to the concepts of multiplicity, becoming, and affirmation, concepts that will be important themes in this book.

## 3 Van Riessen

Again, although not a direct influence on Deleuze, van Riessen is an important philosopher to consider. He was initially an electronic engineer who later became a student of Herman Dooyeweerd (1894–1977), a Dutch philosopher of technology and one of the founders of reformational philosophy, a philosophy grounded in a Calvinist Christian doctrine. Influenced by Mitcham (1994) in his book *Thinking through Technology*, de Vries (2010) discusses the four modes of technology as proffered by Mitcham: technology as objects or artifacts; technology as knowledge; technology as activity and technology as volition (Mitcham, 1994). De Vries argues that philosophy of technology as delivered in the Continental style, tends to orientate towards the last of Mitcham's four modes: technology as volition. The first three modes, de Vries goes on, were largely unexplored until Dooyeweerd, followed later by van Riessen, developed a more analytical philosophy of technology, one that "opened up the black box of modern technology" (Mitcham, 2010. p. 12).

Van Riessen took the work of Dooyeweerd to a new level. For example, he examined the Dooyeweerdian differentiation between 'encapsulation' and 'part-whole' relationships and reconceptualised them. De Vries explains it thus:

> In the case of [encapsulation], a part loses its meaning, once detached from the larger whole, while [in a part-whole relationship] it can still make sense. The screws and bolts in a lathe can be used in other devices, but other, more specific parts cannot. The screws and bolts are related to the lathe in a part-whole relationship, but the sledge is encapsulated in the lathe. Van Riessen recognised that a major development in machine building was the shift from encapsulation to part-whole relations in the context of standardisation and mass production. He called this 'neutralising function division.' He saw this as one of the main characteristics

> of modern technology, compared to traditional craft-based technology. (de Vries, 2010, p. 5)

Van Riessen went on to differentiate between knowledge in science and knowledge in technology, "The former is analytic, abstract and universal, while the latter is synthetic, concrete and specific" (de Vries, 2010, p. 6). However, van Riessen, like Heidegger, also differentiated between old craft-based technology and modern technology, in which the modern begins a progression towards the merging of science and technology into what is now generally accepted as 'technoscience.' Van Riessen thus, argued that the evolution of technoscience coincided synchronously with the evolution of mass production. This, he thought, would result in "a loss and as the cause of many problems concerning the influence of [mass technoscientific development] on culture and society, such as a feeling of estrangement [...] by many people" (de Vries, 2010, p. 6). Developing his argument further, van Riessen mapped his somewhat prescient cultural critique of technoscientific development onto the politics of power. In this respect technology can be seen as a tool to control societies, much as we are experiencing today with digital technologies in particular.

Van Riessen's philosophy, up to this point, can be interpreted as having some similarities to certain aspects of the philosophy of Gilles Deleuze. Whilst I concede that Deleuze is a poststructuralist philosopher and so is linked to the Continental school of philosophy, he is nevertheless considered to be a philosopher of difference. I would contend that van Riessen differentiates between science and technology, craft-based technology and modern technology and power and disempowerment. These are all themes that find some degree of synchronicity with aspects of the work of Deleuze. However, the major disjunction between reformational philosophy and the philosophy of Deleuze, for me, appears to be firmly embedded in the distinction between the concepts of transcendence and immanence. Both terms are relational but constitute a different mode of metaphysics. Transcendence is a relation 'to' something, immanence is a relation 'in' something. Reformational philosophy is Christian in nature and so relates 'to' a God who transcends humankind. Immanent philosophy rejects the idea of transcendence, preferring instead to follow the doctrine of univocity as proposed by Spinoza. "This leads to an interesting concern in Deleuze with notions of eternity that resist definitions in terms of transcendence. We are not immortal in the way we can rise to a different realm (of God or of Platonic Ideas), but in the way we participate in eternal processes" (Williams, 2010, p. 129).

Simondon's work also resonates to some extent with van Riessen, in that it echoes, to some degree, van Riessen's analysis of a lathe. Simondon uses the example of a car engine to explain his philosophy of the differences between what he terms the abstract and the concrete object:

> In a contemporary engine, each important item is so well connected to the others via reciprocal exchanges of energy that it cannot be anything other than what it is. The shape of the combustion chamber, the shape and size of the valves, and the shape of the pistons all belong to the same system within which a multitude of reciprocal causalities exist. To such a shape of these elements corresponds a certain compression ratio, which in turn requires a determinate ignition timing; the shape of the cylinder head, as well as the metal it is made of, produce a certain temperature in the spark plug electronics in relation to all the other elements of the cycle; this temperature in turn causes a reaction leading to the characteristics of ignition and hence to the entire cycle. One could say that the contemporary engine is a concrete engine, whereas the old engine is an abstract engine. In the old engine each element intervenes at a certain moment in the cycle, and then is expected no longer to act upon the other elements; the pieces of the engine are like people who work together, each in their own turn, but who do not know each other. (Simondon, 2017, pp. 26–27)

It becomes apparent for de Vries (2008) and Verkerk et al. (2016) that in attempting to characterise the nature of technological artifacts, Simondon presaged the research programme Dual Nature of Technical Artefacts that was carried out by Kroes and Meijers from 2000–2004 at Delft University of Technology. This project argues that

> technical artefacts have a dual nature: on the one hand they are physical objects (man-made constructions) that may be used to perform a certain function, on the other hand they are intentional objects since it is the function of a technical artefact that distinguishes it from physical (natural) objects and this function has meaning only within a context of intentional human action. (Kroes, 2002, p. 291)

The Dual Nature approach also exhibits some correspondence with the theory of modal aspects developed by Verkerk et al. (2016, p. 83), also at Delft:

> First, both approaches distinguish between different aspects or characteristics. The theory of modal aspects distinguishes 15 different dimensions and the dual nature approach distinguishes between structural and functional characteristics. The subject functions of the artefact (that are

> only present in the lower aspects) relate especially to the physical nature of an artefact and the functional nature is particularly determined by the object functions of the artefact in the higher aspects.

What this demonstrates, at least to some degree, is that philosophical perspectives do not operate in a vacuum. They are subject to interpretation, reinterpretation and supplementation by variations in human thought. Subjectification is thus, a way of "inventing new possibilities of life, [...] to establishing what one may truly call styles of life: here is a vitalism rooted in aesthetics" (Deleuze, 1995, p. 91). Suffice to say that whether from a perspective orientated towards a transcendental mode of existence, as argued by Verkerk et al. (2016), or one of immanence, as argued by Deleuze, it is important to clarify one's position in any philosophical context, indeed in any context in which ideas differ, by recognising, as Dooyeweerd acknowledges, that:

> [p]eople think and act from the perspective of certain views about the world ... Therefore even in the theoretical way of knowing, worldviewish or religious controversies can occur. These cannot be solved by reproaching one another for worldviewish or theological dogmatism but solely by being, as a scientist, as explicit as possible about one's own worldviewish or religious points of departure. (Verkerk et al., 2016, pp. 318–319)

It is in this spirit that I hope to write this book and can equate this line of thinking to Nietzsche's concept of perspectivism which will be discussed later.

## 4 Bernard Stiegler

Finally I want to discuss the philosopher Bernard Stiegler who was very much influenced by Deleuze. He was born on the 1 April 1952, a French philosopher of technology, although his more recent work has become more politically and culturally orientated. He was head of the *Institut de recherche et d'innovation* (IRI), which he founded in 2006 at the Centre Georges-Pompidou. He was also the founder in 2005 of the political and cultural group, *Ars Industrialis*, and the founder in 2010 of the philosophy school, pharmakon.fr, held at Épineuil-le Fleuriel. Arguably, one of his best-known work is *Technics and Time: The Fault of Epimetheus.* This is the book that I concentrate upon and forms a serious influence on the writing of this book. It is in this book that he makes the claim, one that I very much support, that Technics most properly should be thought of as the key philosophical question of our time (Stiegler, 1998, p. 10).

More than eighty years ago, Oswald Spengler (Spengler, 2015) suggested it was hands. Today, Bernard Stiegler (1998) thinks it was feet. Whether hands, feet or both, anthropological evidence, coupled with ongoing philosophical debate, tends to support the idea that the evolution of the human being was, and continues to be, a direct result of its relationship with technology, and, significantly, vice versa; the evolution of technology was, and continues to be, a direct result of its relationship with human beings. Whilst this relationship is undoubtedly complex, as all relationships tend to be, biological human evolution and adaptation, in accordance with technological evolution and adaptation, have, according to Stiegler, always been co-evolutional. This is contrary to Spengler (2015), who argues that the human form came about "suddenly [and] in its entirety" (p. 42). Thus, for Spengler, the completely formed human, together with "hand, gait, and posture [came] into existence together, but – and this is a point that no one hitherto has observed – hand and tool also" (p. 42). Spengler goes on to suggest that whereas the human was suddenly formed in its entirety, there was an evolutionary process that articulated the formation of the hand in association with the tool or weapon: "As the implements took form from the shape of the hand, so also the shape of the hand from the shape of the tool" (p. 43). The concept of a completely formed hand being in any way active without being in relation to technological implements in the form of tools, is an anathema to Spengler. So, for Spengler, whilst the human form evolved suddenly and completely, the hands alone evolved in association with tools.

Stiegler agrees with this up to a point. However, he parts company with Spengler on two significant issues: First, he argues that the evolution of feet preceded that of hands. In the human "everything begins with feet" (Stiegler, 1998, p. 148), and second, he disagrees that the human was fully formed before entering into a relationship with technology. Indeed, he suggests quite the reverse; technology is "The Invention of the Human" (Stiegler, 1998, pp. 134–179).

## 4.1 *The Myth of Prometheus*

In the first of his three books on Technics and Time, Bernard Stiegler discusses the myth of Prometheus. Using it as an analogy to set the scene, he postulates that the evolution of the human being and technology, was, and continues to be co-evolutional. Most of those who study the philosophy of technology will be cognisant of Prometheus, who, according to the myth, was responsible for giving human beings fire and technology. However, many are not aware that Prometheus had a brother named Epimetheus, whose story, according to

Stiegler, is vital in being able to form an understanding of the origins of the relationship between technology and humans. It is Protagoras who narrates the myth and I have reproduced it below.

> Once upon a time, there existed gods but no mortal creatures. When the appointed time came for these also to be born, the gods formed them within the earth out of a mixture of earth and fire and the substances which are compounded from earth and fire. And when they were ready to bring them to the light, they charged Prometheus and Epimetheus with the task of equipping them and allotting suitable powers to each kind. Now Epimetheus begged Prometheus to allow him to do the distribution himself – "and when I have done it," he said, "you can review it." So he persuaded him and set to work. In his allotment he gave some creatures strength without speed and equipped the weaker kinds with speed. Some he armed with weapons, while to the unarmed he gave some other faculty and so contrived means for their preservation. To those that he endowed with smallness, he granted winged flight or a dwelling underground; to those which he increased in stature, their size itself was a protection. Thus he made his whole distribution on a principle of compensation, being careful by these devices that no species should be destroyed ... Now Epimetheus was not a particularly clever person, and before he realized it he had used up all the available powers on the brute beasts, and being left with the human race on his hands unprovided for, did not know what to do with them. While he was puzzling about this, Prometheus came to inspect the work, and found the other animals well off for everything. But man, naked, unshod, unbedded, and unarmed, and already the appointed day had come, when man was to emerge from within the earth into the daylight. Prometheus therefore, being at a loss to provide any means of salvation for man, stole from Hephaestus and Athena the gift of skill in the arts together with fire – for without fire there was no means for anyone to possess or use this skill – and bestowed it upon man. In this way man acquired sufficient resources to keep himself alive ... through this gift man had the means of life, but Prometheus, so the story says, thanks to Epimetheus, had later to stand his trial for theft. Since, then, man had ... by the art which they possessed, soon discovered speech and names, and invented houses and clothes and shoes and bedding and got food from the earth. (Plato in Stiegler, 1998, pp. 187–188)

Stiegler uses and develops this myth to demonstrate a number of issues. One issue considers the question 'what is a human being?' In modern language the actual meaning in ancient Greek of Epimetheus is "afterthought," whereas Prometheus means "forethought." A synonym for afterthought is "forgotten." Stiegler highlights the fact that whilst Epimetheus furnished all other creatures with essential qualities, qualities that gave some defining essence that formed part of their DNA, he forgot humans. Humans, thus, lack any essential quality that answers the question 'what is a human being?' Prior to Prometheus' intervention, humans lacked any essence whatsoever. They were animals without any essential qualities; animals in the raw so to speak. Stiegler refers to this state as "une panne d'essence, a 'lack of fuel,' an 'empty tank'" (Stiegler, 1998, p. 121). The origin of the human is, according to this myth, a default of human origin as being forgotten, as lack. This lack must, therefore, be supplemented from some other source.

It is the forethought of Prometheus, with the help of Athena, that finally resolves this question of lack, up to a point. When Prometheus bestowed the gift of fire and Athena gave skill in the arts, they endowed humans with an essential quality that Stiegler calls technics. Unlike the internalised essential qualities given to all other animal species, technics is a quality that is external to humans. It is not part of their DNA. It is an external technical gift of prosthesis that enables the human to survive and adapt, not only to their environment, but ultimately, by adapting and reforming their environment to suit their needs.

However, the myth of Prometheus offers only a partial perspective on the origin of the relationship between technology and human. Humans, in the myth, follow the ideas conceptualised by Spengler and many others including Rousseau, who is subject to a major critique by Stiegler in Technics and Time 1. For Stiegler, the issue of discordance relates to the idea promulgated by both Spengler and Rousseau that the human came into being fully formed, a hypothesis that Stiegler rejects.

### 4.2 *Criteria of Humanity*

Whilst Stiegler is very much influenced by Simondon and Deleuze, he also draws heavily upon the work of Andre Leroi-Gourhan, a French anthropologist, paleontologist and philosopher. Leroi-Gourhan supports Stiegler's thesis against that offered by Rousseau who claims that:

> Comparative anatomy has as yet made too little progress … to afford an adequate basis for any solid reasoning … I shall suppose his [the human being's] confirmation to have been at all times what it appears to us at this day; that he always walked on two legs and made use of his hands as we do. (Rousseau, 1973, p. 52)

In contrast, Leroi-Gourhan and Stiegler offer a "criteria of humanity," the evolution of which occurred over a vast period of time. The first, and most important criteria, is upright posture made possible by the development the feet. This in turn led to the freeing, and thus the development, of the hands. Upright posture also facilitated the development of language and writing. None of these given criteria, however, could have been made manifest without the co-evolution of technology. Eric Wulf, who was a highly regarded anthropologist, offers a succinct account of the development espoused by Leroi-Gourhan and subsequently by Stiegler

> [T]he relationship between hands and words as conditioned by the structure of the human body is extremely important for the development of language. When humans began to walk upright, thus freeing the hands from their role in locomotion, this also allowed them to use their hands to grip objects. The next step was the handling of foodstuffs and then, gradually, the use and manufacture of tools. As humans began using their hand to grip objects, they no longer needed to use the mouth for this task. In the same way as the hands, once liberated, progressed to the use and manufacture of tools, the face now developed methods of symbolization, which, as they were articulated, turned into a system for producing sounds. These two developments are both part of one and the same process. "As soon as there are prehistoric tools, there is the possibility of language, for tools and language are neurologically linked and cannot be disassociated within the social structure of humankind" (Leroi-Gourhan, 1993, p. 114). In common with all primates, human beings also have a neuronal connection between the hand and facial organs. However, in contrast to other primates, humans, with their upright gait, can create symbols and tools. To put it another way, humans, though they started out with the same formula as primates, can make tools as well as symbols, both of which derive from the same process, or, rather, draw upon the same basic equipment in the brain. This leads us to conclude not only that language is as characteristic for humans as our tools, but also that both are the expression of the same intrinsically human property. (Wulf, 2013, p. 239)

Leroi-Gourhan (1993) thus argues that the formation of tool fabrication and the co-development of language, ultimately resulted in the formation of psychomotor skills, intellectual and inventive capacity as a direct result of the process of erect posture. This is in distinct opposition to the theory proposed by Rousseau, in which these capacities were already inherent within the fully evolved human who came into being ready made. In other words, for Rousseau, these essential qualities were genetically endowed as part of the process of human evolution and were independent of the technological potentialities that surrounded it. The fully formed human subject was then able to act independently upon the natural objective world, by inventing and fabricating technologies by wit alone. According to this theory, the qualities endowed upon the human, just as in other animal species, are internal, as are the internal qualities of speed to the leopard or flight to the bird. Leroi-Gourhan and Stiegler disagree with this interpretation. They make a clear distinction between Rousseau's perspective and their own.

The philosophers mentioned all have one thing in common. No matter how the human came into existence, the human relationship with technology has evolved co-terminously, from an initial phase that might be described as primitive, to the phase that we encounter today. The natural world has been technologised by human beings, and this has occurred continually over the entire period of human existence. We have now reached the era referred to as the Anthropocene. This book will argue that since these tectonic technological phase changes are becoming an existential threat to society, there is an urgent need for a forum or Agora in schools in which young people's voices can be heard.

## 5 Deleuze, Guattari and Simondon on the Concept of Becoming

We have a tendency to see the world in stable terms. Attempts to deliver technological literacy serve as examples of this. The ITEA Standards (2000, 2002, 2007) "have gone as far as developing a series of standards that spell out in detail how to achieve technological literacy and what goals need to be achieved at different grades within a school curriculum" (Ingerman & Collier-Reed, 2011, p. 138). Thinking this way implies that one, with some degree of effort, can, by means of examination, become, for example, a level A, B, or C, or, indeed fail as a technologically literate human being. Deleuze and Guattari, however, reject this notion and offer instead a philosophy of the becoming of being. This is an important concept to review. It enables us to differentiate between technological literacy as a subject area which requires a preordained set of standards that can be assessed quantitively, and (ethno)technological literacy as a subject area that is qualitative in nature and so must be formatively assessed.

Ilya Prigogine, winner of the Nobel Prize in 1977 for his work on the thermodynamics of nonequilibrium systems, co-wrote, in association with his colleague Isabell Stengers, an acclaimed philosopher, a pioneering book entitled *Order out of Chaos* in 1984. The philosophy behind this book challenged the thinking behind the 17th and 18th century versions of classical science, or Newtonianism as it was then known. Newtonianism characterises thinking as offering a deterministic view of nature as being in a state of permanence; one that could be studied, broken down and predicted with certainty, thereby implying nature as being fixed and stable (Toffler in Prigogene & Stengers, 1986, p. XIII). Prigogene and Stengers challenged this line of thinking as follows:

> Science certainly involves manipulating nature, but it is also an attempt to understand it, to dig deeper into questions that have been asked generation after generation. One of these questions runs like a leitmotiv, almost as an obsession, … through the history of science and philosophy. This is the question of the relation between being and becoming, between permanence and change. (Prigogene & Stengers, 1986, p. 291)

Gilles Deleuze is considered to be a philosopher of difference or change in the form of continually becoming something other. He has written many books about other philosophers including Hume, Spinoza, Bergson, and Nietzsche. However, it was not until 1968 that he published *Difference and Repetition*, the first book to outline his own unique, and it must be said complex philosophy. In his own words "this was the book that I tried to 'do philosophy.' All that I have done since is connected to this book" (Deleuze, 2004a, p. XIII). Many subsequent philosophers consider this book to be his magnum opus. Inspired by Bergson, Deleuze outlines two kinds of difference: differences in kind and differences in degree, as Elizabeth Grosz (2004) explains:

> Differences in degree, which he construes as differences of magnitude, quantitative differences, differences of more or less, measurable differences; and differences of nature or kind, which are qualitative differences, differences impossible to measure or describe in numerical terms but discernible in and for conscious mental life and experienced in the continuity and ever-changing movement of duration. If quantitative differences are measurable, they indicate spatial differences, differences external to each other, differences between things, *differences* that can be marked or characterised through their coordination with a third term, a number, that is, through measurement. Such differences are discrete, discontinuous, and homogeneous. They can be divided in infinite ways without transforming their nature. (pp. 158–159)

Deleuze denotes quantitative differences or multiplicities as being extensive; they are actual, objective and possess an identity (cars for example). Qualitative differences or multiplicities, on the other hand, are continuous, subjective, are experienced in lived time and differ in kind from one another. They can be thought of in terms of human moods, emotions or ideas (Tampio, 2010).

As human beings, we are deeply anchored into thinking of difference in quantitative, measurable terms. 'The 'Shard' is the tallest building in London. It stands 309.6 meters high.' 'I got '82%' in my maths test. In my last maths test, I only achieved a score of 81%,' etc. But, in fact, we think of difference in qualitative terms, albeit less consciously, much more than we do in quantitative terms. We just do not necessarily realise it all the time. We may rarely ask questions such as: why is knowing that the Shard is the tallest building in London of interest? How does a score of 82% in your maths test affect you? Whilst we may rarely ask these types of qualitatively orientated questions, however, we are, nevertheless, affected by them. And it is here that we can begin to differentiate between the concepts of being and becoming. Being the tallest building in London is a label that whilst true, is not a permanent state, it is a semi-permanent state or as Simondon proposes, a multistable state. It is likely that another taller building will be constructed later, one that supersedes the height of the Shard, thus relegating it to becoming the second tallest building in London. Being a student who got 82% in a maths test is also not likely to be a permanent state. Other tests in the future will probably change the status of the student to a higher-grade student in maths or, perhaps a lower-grade one. All quantitative scenarios depict a state of being bigger, better or different by way of measurement. In terms of a human being, difference can also be measured in both quantitative and qualitative ways: I am a male and my wife is a female; I am taller than my wife; I am fatter than my wife. These are quantitative differences that are compared by physical means, differences in degree between two people. Another example: After leaving school, I began my career in the field of architecture but then, some years later, I changed my profession and became a technology education teacher. The difference here is quantitative in that I have moved between two quantitively different professions, one to another. This is a difference in degree signifying differences between two occupations. However, these differences in degree offer only a partial understanding of this particular event in my life. It can be easily verified that I changed from one occupation to another through tax records or testament from former colleagues, but that is only knowing *that* I changed occupations. This is declarative knowledge. It also, in the telling, presupposes that I had somehow gained the requisite skill set that enabled me to know how to teach technology education. This is both declarative (some form of certification) and procedural (know how to) knowledge. These are all differences that are made manifest in a way that can be judged objectively by some external agent. When compared to my wife, it is obvious that I am taller and fatter. I have a certificate that verifies my qualification to teach. What is not made manifest, what is not made clear, is why? What were my reasons for changing? These are, as Grosz describes above, akin to differences of nature or kind, differences that are impossible to measure or describe in numerical terms but are only discernible to me in my experiences within the continuity and ever-changing movement of duration. They are differences that, whilst internal to me, will have been both effected and affected by external forces.

Later still, I became a university lecturer studying the philosophy of technology and education, which is yet another quantitative-qualitative difference. The three changes that I have reported represent a significant, indeed intensive set of differences that affected not only me, but also many others (students for example). These three stages represent only a microcosm of the rich and varied transformative differences that happened over one part my life. The differences that occurred as part of the process of my transition from technology teacher to university lecturer, came into being as a result of the many different forces I came into contact with, forces that battle amongst each other on many complex fronts. These forces, many of which cannot be foreseen, eventually produce an event, which in my case, resulted in the transition from being a technology teacher to becoming a university lecturer. Difference in this respect is the becoming of a new being which in turn will differentiate into becoming yet another new being and so on. Figure 3 represents the quantitative difference in my transition which is easy to both define and comprehend. I was a technology teacher, a clear and distinct occupation. I then became a university lecturer, another distinct occupation.

However, Figure 3 suggests that my move from being a technology teacher to becoming a university lecturer constitutes two completely different, distinct and independent states of being. This would mean that there must be an actual break, a rupture from one state of being to the next: after all, I cannot be both at the same time. This now raises several questions such as, 'why did I change from one to the other?' 'How did I change from one to the other?' 'What happened in the liminal space between my being a technology teacher and being

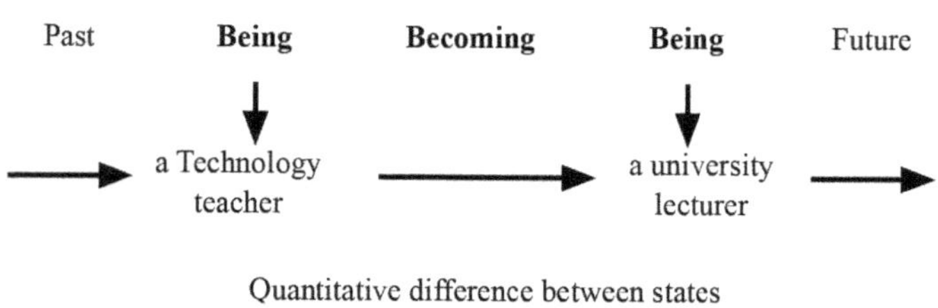


FIGURE 3 Quantitative difference

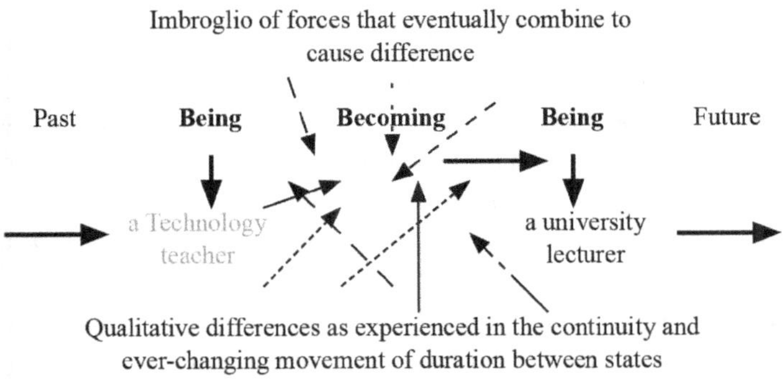


FIGURE 4 Qualitative difference

a university lecturer?' Answers to these questions cannot be measured as they are qualitative in nature.

Figure 4 illustrates the qualitative differences that take place from one phase state to the next. They do not take place in the sense that I stop being one and then start to be the other. Clearly, this whole process occurs over time and involves an imbroglio of chaotic forces that cross between each state of being. For example: 'did I make the right decision?' 'Will this involve having to move house?' 'Do I really want to leave teaching?' and so forth. As I transition from technology teacher to becoming university lecturer, my former self fades into memory as my new self emerges. It is still me, just a different version of me, one that is likely to change again in the future when brought about by other dynamic sets of forces that will continue to influence the evolution of my life.

What is important to note is that my state of being changes as a result of the differences between phase states. Phase states are not static forms of being; they are continuous dynamic processes of becoming, sometimes intensive becomings, sometimes minor becomings. When I became a teacher, I continued my journey towards other becomings, although I will not know how any of these becomings might manifest themselves until the multitude of ongoing forces that direct my life cause many events to occur in many possible directions. Life involves an ongoing process of change: nothing is permanent. Indeed, thermodynamics tells us that permanence is either an impossibility or will result in death. Being anything is a metastable state. Permanence is an illusion. Craft-based technology education tends to favour the quantitative over the qualitative. Ethnotechnological literacy reverses this dynamic.

For Simondon and Deleuze, the enduring state of being 'some-thing' is a concept that they reject. For them, no-thing in the universe is stable and enduring; all things are multistable. I discussed the becoming of a human above, but for these philosophers, every-thing, including the human, is subject to many phase changes or modes of existence in their existential being over time. They are continually in a state of impermanence, or in Souriau's terms 'existential incompleteness' (Souriau, 2015). Whether natural or artificial (or both), all beings or 'things' become something different from that which they were before. Every-thing is an evolving process. They become things that undergo change in some way, change that affects them and they in turn affect other beings as part of that process. This all happens in a continuous series of phase changes that occur within events. They do not occur in a space that distinguishes two distinct and identifiable events from each other and which thereby separates one set of events from the next. Rather, they develop as a continuation of changes that flow through events, thereby altering the course of their production (Stagoll, 2010a, p. 26).

## 6 The Becoming of Being

In his book *The Logic of Sense*, Deleuze explains why time does not follow a determinable, thus linear path, a path that can be retraced from present to past. This is because universe acts in a chaotic and unpredictable manner. In the chapter entitled "Porcelain and Volcano," Deleuze analyses F. Scott Fitzgerald's confessional memoir "The Crack Up." In this, Fitzgerald describes the process he went through, from being a handsome, wealthy and successful novelist in his late twenties, to becoming a broken down, penniless alcoholic in his late forties. Fitzgerald attempts to pinpoint exactly when his 'crack up' occurred and in so doing, seeks to reconstitute his life back to the way things were before the change happened and thus revert to what he then considered normal (Deleuze, 2004a, pp. 176–183; Butler, 2016, p. 185). But of-course this is not possible, because even if memory allowed it, all the events that happened between the past event and the present, could not be recalled sequentially. This is because any defining set of events that occurred in the past could only ever be the result of many other converging forces, both quantitative and qualitative. Many of these, which are unknowable, may or may not have served to influence Fitzgerald's trajectory into alcoholism. Thus, it is impossible to reform a life anew, exactly as it was at some time in the past. All the intervening forces that occurred over the time between then and now, will have influenced that life and, in the process, made it a different version of itself. Life can flow only in one direction. As Heraclitus suggests "*it is always something new that is reborn and not the same entity as before*" ... "one cannot step twice into the same river, nor can one grasp any mortal substance in a stable condition, but by the intensity and the rapidity of change it scatters and again gathers" (Kahn, 2004, pp. 168, 222, emphasis added).

Considered in contemporary terms, the Covid pandemic has everyone wishing things would 'get *back* to normal.' However, what was normal pre-pandemic cannot be reconstituted post-pandemic: that would be akin to wishing, in actuality, for the entire world to go back in time. This is clearly impossible. Everything changes as time changes and things cannot be reconstituted in actuality but only by way of reproduction as in, for example, movies or museums. Time for Deleuze and Guattari:

> is non-linear and irreversible. Reversible, linear time corresponds to the mechanistic, calculable view of the universe made famous by Newton: two billiard balls on a collision course always interact the same way, and if you could rewind and replay the interaction 100 times, the billiard balls would take the same trajectories every time. Start the process of evolution over 100 times, however, and you would get up to 100 different results: this is an example of the difference between linear-mechanistic causality and non-linear, emergent causality; the latter involves singularities or bifurcation-points, and it is particularly at these undecidable points that time reveals itself to be irreversible. (Holland, 2003, p. 27)

Likewise, Elizabeth Grosz supports the idea that the process of production of an individual cannot be fully comprehended in Newtonian terms. She explains this in Simondonian terms:

> [Simondon] aims to avoid the usual processes of reverse engineering, in which a given object's process of production is deduced from the identity of the constituted object in the present. Such a process can only move from identity to identity, from one individual to all those that precede it. (Grosz, 2012, p. 38)

What Grosz is saying is that, for Simondon and Deleuze, the genesis of the production of an individual, or object, is much more complicated than a simple singular time-line as shown in Figure 5.

The type of model, shown in Figure 5, makes the assumption that an individual has a series of fixed, stable and separate identities that can simply be traced back to their origin by reverse-tracing a series of separate and uniquely identifiable events. Each version would be hermetically sealed within a bounded system, thereby retaining its unique identity by separating it from all other 'different' versions and forces.

Simondon and Deleuze consider this to be a false perspective. Any bounded system, for them, must be porous in order to allow the myriad of external

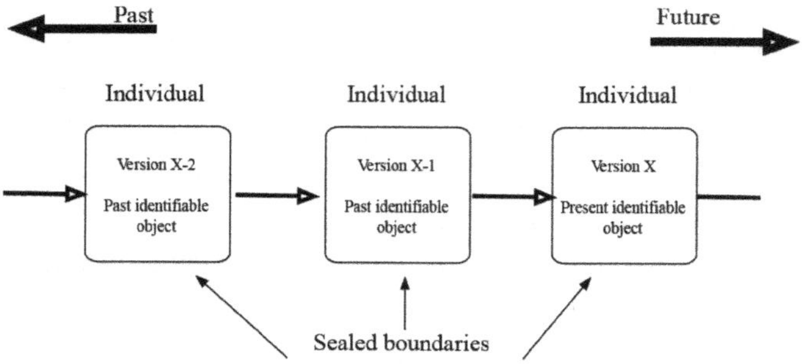


FIGURE 5 Graphical depiction of a linear time-line

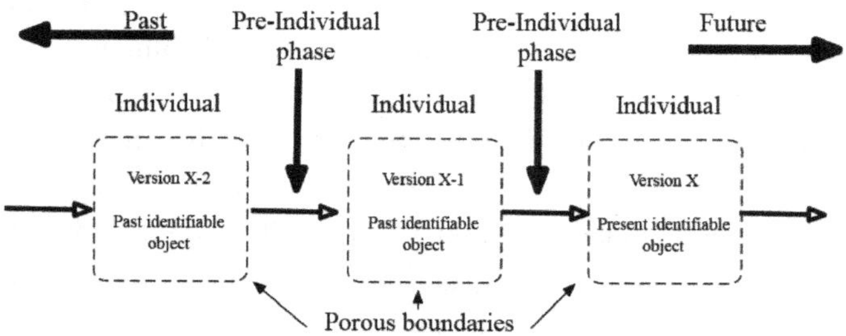


FIGURE 6 Graphical representation of the pre-individual phase

forces that exist to permeate the system, as part of the process of creating a new individuated being. This process is always prior to the actualisation of any novel version of an individuated being. It is the process of the becoming of being. Both Simondon and Deleuze and Guattari introduce their concept of 'individuation' in this respect. This is where any actualised identifiable object, whether organic or non-organic, goes through a prior process of pre-individuation before it is made manifest as an actual individual being. This process is in a state of continuous flux, going from one phase state to another. Between each concrete actualisation – becoming individual – is, for Deleuze, a virtual phase. Simondon refers to it as the 'pre-individual' phase (see Figure 6).

It must be noted that the pre-individual phase does not reside between any two concrete actualisations. Any technological individual is subject to forces of change as soon as it becomes actualised. It goes immediately from being individuated with an identity, to becoming other (for Deleuze), or going through a process of pre-individuation (for Simondon), a phase where its identity is subject to forces of change until is it once more actualised into a new conceptualisation of its former self. For technical objects, unlike humans, the previous version remains intact, which is why archaeologists are able to speculate upon the past by analysing the hidden technologies they find. The process of

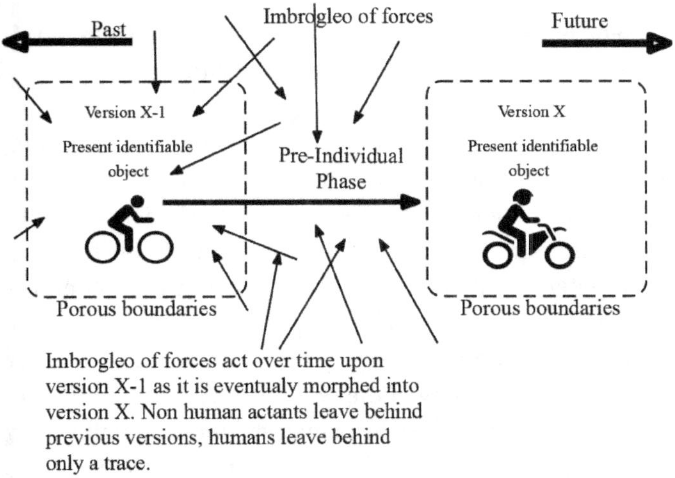


FIGURE 7 The becoming of being or the pre-individual stage

becoming for a human, on the other hand, leaves only a trace of the former self, a memory of that phase of being.

As a technological example, consider the becoming or individuation of a motor bike. It is born out of its forebear, the bicycle. The bicycle will have gone through many iterations before becoming a motor bike. Lots of forces will have contributed to its incarnation: the development of ideas, new materials, new techniques, petrol engines, pneumatic tyres; efficient brake systems; gears; handlebar throttles and so on. I can offer only an extremely limited account of the multitude of forces that came into play during the individuation of the motor bike, having no way of knowing what lay ahead. This is why in any process of the becoming of being or individuation, it would have been impossible to foresee all of the forces that would come together to influence the next phase. It is also impossible to trace these forces backwards. Figure 7 demonstrates the chaos involved in what Deleuze refers to as the becoming of being and Simondon calls the pre-individual phase.

## 7 The Concept of Assemblages or Multiplicities

In this section, I want to reinforce the discussions above on becoming, by examining both Deleuze's concepts of assemblage and Simondon's concept of individuation. An assemblage, for Deleuze, is that of a multiplicity, a metaphysics in which the concept of multiplicity replaces that of substance, in which event replaces essence and virtuality replaces possibility (Smith & Protevi, 2020).

One of the initial influences for Deleuze in this respect was the calculus developed originally by Gauss and later expanded upon by Riemann.[1]

> Mathematics played a major role in Gilles Deleuze's thought, beginning with his engagement with calculus and Gottfried Leibniz, who was also a major philosophical influence on Deleuze. Bernhard Riemann may, however, be the most significant mathematical influence on Deleuze, especially in … his [later] collaborations with Félix Guattari. The conjunction of Riemann's mathematics and Deleuze's philosophy is a remarkable event in the history of twentieth-century philosophy, and it has major implications for our understanding of the relationships between mathematics and Deleuze's thought, and between mathematics and philosophy in general …. Deleuze, I would argue, takes advantage of both Riemann's mathematical and philosophical concepts in building his own philosophical concepts. (Plotnisky, 2009, p. 190)

The mathematical concept of Calculus was developed originally by Newton and Leibniz, although quite separately, and presented by both pretty much simultaneously. Arguments continue as to who exactly was first, but that is of little importance to this book. The calculus enabled a dynamic form of a mathematics of continuous change, a form of mathematics that had not existed before. This built upon the work of Descartes. Descartes, who was also a highly regarded mathematician, invented Cartesian graphs. His 2D x, y graphs and 3D x, y, z graphs are well known to most secondary school pupils. These graphs plot a 2D or 3D shape in a fixed relation to the external coordinates of the graph. If you know the coordinates, you can replicate the shape an infinite number of times. The coordinates represent a hierarchical or transcendental state. They bind the shape to its externally bounded coordinates, thereby designating a fixed and stable state that guarantees repeatability. Any change to the shape results in a completely new independent shape with its own separate, fixed and repeatable identity. The calculus of Gauss and Riemann offered a new system that allowed for continuous change in the shape or in the whole, but one that did not rely upon external fixed coordinates.

It is not my intention to discuss Riemann mathematics any further since that is beyond the scope of this book. Suffice to say that Deleuze was influenced by Riemannian mathematics when he first postulated the idea of an assemblage. This is something he later refined in collaboration with Guattari.

> Assemblage theory provides a bottom-up framework for analysing social complexity by emphasizing fluidity, exchangeability, and multiple

> functionalities through entities and their connectivity. Assemblage theory asserts that, within a body [human or non-human], the relationships of component parts are not stable and fixed; rather, they can be displaced and replaced within and among other bodies, thus approaching systems through relations of exteriority. (Wikipedia, 2020)

An assemblage is, thus, not constrained by some essential enduring identity. It can and does, however, have boundaries. These boundaries, however, are porous and subject to variation. The example of the transformation from a bicycle to a motor bike in Figure 7, constitutes an assemblage.

Let us also consider the unification of Germany. Prior to its unification, Germany comprised two distinct countries: East and West Germany. These became separated after the war. It literally had a part physical boundary in the shape of a wall. Before the war, the whole country was Nazi Germany. This then divided and later reunified. These phases represent three bounded states with porous borders where identities are subject to variation over time. Universal fixed identities, therefore, no longer apply. Concepts such as country, school, education about technology, student, state, society, and culture, are all assemblages – multiplicities that have no enduring properties but, rather, possess properties that are subject to continuous states of change, redefinition, and reconstitution. "Deleuze takes the idea that [technology, for example] is composed of different [assemblages] that form a kind of patchwork or ensemble without becoming a totality or whole" (Roffe, 2005, p. 176). In other words, there is no such thing as technology, only an ongoing evolution of the concept of technology.

Change is a continuous process of what Deleuze refers to as difference-in-itself: "at the same time it forms and dissolves" (Kahn, 2004, p. 168). Deleuze argues that we must never assume that a pre-existing identity denotes sameness. A river, for Deleuze, is an assemblage, a multiplicity of forces that forms, dissolves and re-forms continuously.

> A multiplicity is, in the most basic sense, a complex structure that does not reference a prior unity. Multiplicities are not parts of a greater whole that have been fragmented, and they cannot be considered manifold expressions of a single concept or transcendent unity. On these grounds, Deleuze opposes the dyad One/Many, in all of its forms, with multiplicity. Further, he insists that the crucial point is to consider multiplicity in its substantive form – a multiplicity – rather than as an adjective – as multiplicity of something. Everything for Deleuze is a multiplicity in this fashion. (Roffe, 2010, p. 182)

There is now emerging a clear connection between the concept of quantitative and qualitative differences, and multiplicities. Tampio demonstrate how Deleuze's theory of multiplicities serves, in many senses, to begin to resolve C. P. Snow's Two Cultures debate between the sciences and humanities (Snow, 1998), and between analytical and continental philosophy. The concept of multiplicity reconfigures the ancient dispute between reason and poetry. Tampio (2010) posits

> Quantitative multiplicities can be captured through logical, mathematical, or scientific propositions. Qualitative multiplicities require a broader palette to colour in the nuance of each thing, and here, philosophy forms alliances with painters, authors, directors, or sculptors. Philosophers create concepts that align with artistic percepts. That is one reason that Deleuze cites artists such as Virginia Wolff or Marcel Proust. When Wolff experiences herself as a school of fish, she portrays the lived sensation of being a qualitative multiplicity open to powers and affects circulating in the universe.[2] Deleuze constructs philosophical concepts, such as the virtual, to describe the source of animal becomings that permeate our subjectivity. For Deleuze, philosophers should investigate the world using the tools of both reason (quantitative multiplicities) and poetry (qualitative multiplicities).

Extensive multiplicities can, thus, be divided into identifiable and measurable parts. An intensive multiplicity, in contradistinction, "cannot be divided up without changing in nature. In other words, any alteration to an intensive multiplicity means a change in its nature – a change in its intensive state. This is important for Deleuze because it means that there is no essence of particular multiplicities which can remain unaffected by encounters with others" (Roffe, 2010, pp. 181–182).

A well-known scientific experiment may help to explain this. It involves two identical beakers, each half filled with water. In one, the water is at ambient room temperature and in the other, the water is at a temperature of 30°C. When the contents of the two beakers are mixed, the volume of water doubles. This can be measured easily and exactly. It can be likened to an extensive multiplicity. The temperature of the water, on the other hand, alters to somewhere between the two original temperatures and continues to change until an equilibrium of temperature is reached. This will be dependent upon a variety of possible forces, such as the ambient room temperature. This can be likened to an intensive multiplicity. Volume is quantitative and can be divided into identifiable, measurable, numerical parts. The temperature of the water is, on the other hand, subject to encounters with other forces such as pressure, heat or cold, which can change its nature across a spectrum from steam to water to ice. In other words, it changes essentially if it is altered in some fashion as a result of forming a relationship with some other multiplicity of forces.

A fear of drowning, for example, may be considered irrational by a competent swimmer, but it is an intensive, actual state that can be experienced by a non-swimmer. Considered thus, assemblages cannot be reduced to a swimmer/non-swimmer dyad. There are many other factors to be taken into account. It is not beyond reason, for example, that a competent swimmer could fear drowning whilst a non-swimmer had no fear. My grandson, when aged 4, could not swim but loved jumping into the pool and scrambling his way back to the side, mostly under the water. Fear and anxiety are not quantifiable multiplicities. They are intensive states that both the swimmer and non-swimmer may have to engage with and are dependent upon the circumstances experienced at any given time, including experiences from the past. Going shopping, as another example, is unlikely to cause any great anxiety to most, unless the normality of shopping undergoes an intensive shift as the Covid 19 virus pandemic (an external force) reshapes society. "There is no universal nature of any intensive multiplicities which can remain unaffected by encounters with other [things]" (Roffe, 2010, p. 183). Assemblages are about the formation of relationships where competing forces vie with one another resulting, ultimately, in something becoming another version of itself.

A river, for example, is a dynamic relationship of a variety of material forms, organic attributes and habits that, as a result of their continuously changing qualities, cannot be determined as a stable, universal entity. The water flows, sometimes vigorously, sometimes gently. Sometimes it dries up altogether. Moreover, it affects other entities: salmon leaping; riversides eroding; towns flooding. These attributes occur within the network of relationships that the river has with other entities, such as weather conditions or fallen trees which change or interrupt the flow. In other words, the river cannot be considered as some pre-existing identifiable entity in and of itself: it is more than that.

For Deleuze, every aspect of reality evidences difference, but not differences of degree:

> rather than theorising how individuals might be grouped, it is more important to explore the specific and unique development or 'becoming' of each individual. The genealogy of an individual lies not in generality or commonality [differences of degree], but in a process of individuation determined by actual and specific differences, multitudinous influences and chance interactions. (Stagoll, 2010a, p. 75)

The non-swimmer mentioned earlier is a human being. However, the particularity of her being in a river learning to swim is a relationship of interiority as well as exteriority. It will be different for her internally from the experience of going shopping, for example, as will be the experience internally of a competent swimmer, albeit from a different set of relationships. The non-swimmer, whilst learning to swim in the river, will form part of the multiplicity that she finds herself in at that particular moment. It will bring together all her previous experiences and thoughts associated with swimming, together with the surrounding environment of the river, the weather, other swimmers, past experiences etc. She can no longer be classified as being a non-swimmer; a stable category. She is 'becoming' swimmer; not becoming 'a' swimmer as that again designates a stable category suggesting a final end-state. Becoming swimmer is a process that continues; it has no terminal state other than death. The non-swimmer may become a very competent swimmer, a competitive swimmer, a teacher of swimming. She may now want a boat or enjoy going on a cruise and so on. A different multiplicity with a different set of circumstances and competing forces will change the relationship she has with swimming. Becoming swimmer requires the formation of a relationship between action (reality, sensory perception) and signs (virtuality, thinking, translation). Put another way, this involves forming relationships between the way we physically sense conjugating our bodies with the water, which is not a natural habitat for humans, and the way we are affected by it over time through repetition[3] (overcoming our fears and anxieties), and the way we conceive the concept of swimming as being desirable, or not (imagining how to swim).

Becoming swimmer, like becoming anything, requires experimentation and so, risk. This involves taking oneself outside one's comfort zone. Deleuze (1995) puts it thus:

> once one steps outside what's been thought before, once one ventures outside what's familiar and reassuring, once one has to invent new concepts for unknown lands, the methods and moral systems break down and thinking becomes … a perilous act, a violence, whose first victim is oneself. (p. 103)

Semetsky and Delpech-Ramey (2011), citing Deleuze, consider this perilous act of thinking as "a power to affect itself, an affect of self on self" that leads to learning from experience and so becoming other (p. 519). But for Deleuze, learning from experience does not, for example, mean learning how to drive an automobile, a skill, which through constant repetition, will enable a novice driver to become a competent driver, providing they work at it hard enough.

That does not constitute thinking from a Deleuzian perspective. Ironically, the more expert a driver or swimmer or carpenter becomes, the less they need to think about it: it appears effortles, automated, almost magical.[4] Achieving expertise in this sense, means to reach a final end state. There is no stage beyond becoming an expert. It is not unlike many policy makers and educationalists who argue that disruptive students stop other students in the class from achieving their 'full potential,' another final state of being. What happens after they reach their *full* potential? I have never met anyone who claims to have reached their full potential. I am not even entirely sure what 'full potential' means, if anything. For Deleuze and Guattari, thinking requires that we "move beyond what we experience so that we can think of new possibilities ... so as to be able to think anew" (Stagoll, 2010b, pp. 53–54). Thinking is about the creation of something that does not yet exist.

It is in this sense that Deleuze seeks to destabilise our thinking, disrupt our faculties and free our senses from established tendencies, in order that we might uncover the differences evident in the lived world, and realise the uniqueness of each moment and each thing (Stagoll, 2010a, pp. 75–76). "The notion of some 'thing in general' can be set aside in favour of one's experience of this thing, here and now" (Stagoll, 2010a, p. 76). Stable identities, thus flow from context to context: She is a consummate shopper but a poor swimmer. One day she may become good at both. She will nevertheless continue becoming shopper and becoming swimmer because to stop becoming is to give up, remain stagnant, conform, have reached her full potential, cease to exist.

We might, therefore, envisage the various purposeful activities or events that exist as modes of existence, to be represented in circular form. Thus "things, processes, events, formations, constellations, in short, matter in all its permutations," return to be recycled, but never as the same concept. Any recycled multiplicity is always different from its original form, although it will be influenced by previous multiplicities, one flowing into the other (Grosz, 2004, p. 142). Multiplicities are, thus, multistable processes that can never be fully realised or identified. The experience of the non-swimmer's relationship with the river, will cause changes to her mode of existence. She may be less fearful of learning to swim, or perhaps more fearful. She may become able to swim breaststroke but none of the other strokes. The river too might undergo a change to its mode of existence. The river-bed may be altered by her frantic thrashing about. Other experiences will cause change yet again, as individual relationships, determined by actual and specific differences, multitudinous influences and chance interactions, occur. Becoming swimmer, becoming shopper or just becoming other is to evolve, to experiment, to encounter novelty, to think new thoughts, to experience. To remain comfortable with what is familiar and reassuring is akin to getting close to entering into a state of entropy, explained by Schrödinger (2013) thus:

> What is the characteristic feature of life? When is a piece of matter said to be alive? When it goes on 'doing something,' moving, exchanging material with its environment, and so forth … When a system that is not alive is isolated or put in a uniform environment, all motion usually comes to a standstill very soon … After that the whole system fades away into a dead, inert lump of matter. (p. 69)

Examples of this are, perhaps, best exemplified by synonyms for the term stagnant, as in stagnant water or a stagnant economy. Such synonyms include: dormant; idle; inactive, lifeless; static; filthy; foul; passive; putrid; stale; unmoving; inert; dead; equilibrium. Antonyms, on the other hand include terms such as: active; alert; energetic; lively; mobile; moving, alive. Schrödinger goes on to offer a way to avoid entropy by introducing the concept of what he termed negative entropy, or negentropy as it has now become known. This is when a system avoids the rapid decline into the inert state of 'equilibrium' by way of metabolising. This involves more than just the digestion of food. It involves many things that appertain to our appetites in general: our tastes; our friends; where we live; the literature we read; how healthy we are; the activities we engage in. Essentially, the various ways in which we live our lives impacts on our becoming different. It forms our character (Nietzsche, 1911).

Thus far, I have mainly considered human beings as assemblages subject to multiple processes of change as determined by both internal and external influences or multiplicities. I want now to focus on artificial constructs or technologies. Take, for example, a house. In conventional thinking, a house is an objective whole that can be identified as a place where people live. A house has actual properties that can be quantified, like walls and floors and a roof. It has designated rooms to live in, to cook in, to bathe in. A house has windows and doors and rainwater pipes. The problem with this description, however, is that it does not offer a clear, logical, mathematical, or scientific definition for a house; one that can reduce the concept of house to some essential quality. It cannot be reduced to a quantitative multiplicity alone. This is because there is no repeatable, measurable or numerical definition that can be ascribed to the concept of a house. There are no external Cartesian coordinates that can be used to provide a universal declaration of what constitutes a house. As soon as we try to offer one, there is always something else; something more; something beyond; something we had not considered before. Twisting Spinoza ever so gently, we simply do not know what a house is capable of becoming in the future. We do know, however, what the concept of a house has been in historical and cultural terms: a cave; an igloo; a tepee; a yurt; a bungalow; an apartment; a cardboard box; the name given to a type of joint in woodworking; the building in which parliamentarians sit (house of commons or house of representatives); a houseboat, to name but a few. But none of these provides us with any universal, enduring identity, some quality that represents the specific essence of a house. All offer only a general conception related to history and culture. A house can be perceived only contingently and in lived time.

Whilst a house can be considered a quantitative assemblage in that it is actual and identifiable as a house, albeit in a contingent sense, it is also a qualitative assemblage. A house is more than the sum of its quantifiable parts: it is a form of expression; a form of intensity that awakens emotions and feelings. A house can offer protection from the elements. It may be considered a sanctuary, an enclosed space offering privacy, warmth, and security. A house may, for some, appear to express the wealth of its owners, or the power of its inhabitants (as in the case of the White House), or again it may express aesthetic beauty for others. A house is capable of both affecting and being affected. It is a synthesis of both extensive and intensive forces, of quantitative and qualitative multiplicities. It is an assemblage.

Put another way, there is no such thing as a house, at least not in any conventional sense. A house is an event, one that is constituted by the interactions between its parts – human beings, extensions, new windows, burst drain-pipes, floor coverings, furniture, lighting. It may be a chateau or a tent and so on. Moreover, no two assemblages are ever the same. No two houses are ever the same; they are always in a state of constant emergence: different inhabitants, different floor coverings, different furniture, and so on. There is only becoming house, not becoming a house. For Deleuze, things and states are both products of becoming.

Deleuze further makes it clear that assemblages are never merely technological. Any technological elements that are part of an assemblage are always social first and so require human beings to be engaged in the pre-individual phase, or design phase.

> Thus, technology becomes one way of solving the problem of building society on a larger scale. In this sense, even modern technology is social. It represents a further resource in the mobilisation of individuals in the performation of society. (Strum & Latour, 1999, p. 123)

It is the relationship between what Latour refers to as the actants that makes any technological element social (Latour, 2005). Simondon uses the term individualisation as being part of what he refers to as an associated milieu, something which explains the process of the actualisation of a technology (Simondon, 2017). It is in this respect that I will later conflate the concept of Deleuze's assemblage and Simondon's associated milieu.

To summarise, the component parts of an assemblage have histories. These histories are the result of past assemblages or events that are, in turn, the result of past assemblages or events and so on. Any change that may be constituted as a result of the formation of an assemblage can never be fully known in advance. This means that there are multiple potential pathways or bifurcations that an emerging assemblage may take or may have taken in the past. In other words, while perhaps only one emergent pathway may have been actualised in the past, or may be actualised in the future, other potential futures that will have existed in the past may emerge to influence the future. Moreover, anything emerging from an assemblage, whether organic or inorganic, will enter into a new assemblage, or what Simondon calls an associated milieu. It will then be subject to change. de Vries offers a good example of this in the design and subsequent use of a screwdriver. "A screwdriver has been designed to turn screws and in many cases, it will be used that way. But no one can stop the user [in a new assemblage or associated milieu] from using it as a handy device with which tin can lids can be lifted" (de Vries, 2005, p. 16). This is a good example of the quantitative and qualitative forces that occur in any assemblage; the former being the intended use it was designed for, the latter being the unknown user redesignating it for another use, or perhaps another user yet again, redesignating it further, as a weapon. In addition, de Vries suggests that "designers will often not only think about the proper use of the [screwdriver], but also try to anticipate accidental functions and design that the device, in such a way that it can stand all sorts of abuse" (de Vries, 2005, p. 16). In other words, the screwdriver is designed to turn screws. It may, however, break when used as a lever to open tin cans, so this non-anticipated new function may, or may not, influence the design of the next generation of screwdrivers. Think of the evolution of a penknife into becoming a Swiss army officer's knife. These potential futures represent an evolutionary process of becoming other that is chaotic and non-linear.

The scientists who developed the atomic bomb in the Second World War could have refused to do so. They appear to have wished that they did refuse to do so, in retrospect. This suggestion is made manifest in Oppenheimer's famous quote, borrowed from the Hindu sacred text the *Bhagavad-Gita* 'Now I am become Death, the destroyer of worlds.' This suggests an alternative emergent potential pathway that might have been taken, a pathway that would have presented another possible future. In order to understand better why that alternative pathway was not pursued, we would need to examine the various components that came together to form the assemblage or event that we have come to know as a human-made form of mass destruction. Human-designed forms of destruction existed long before the atom bomb and have continued to proliferate ever since. They are constantly upgraded and changed. They too are in a constant state of becoming. Technologies, in whatever form, emerge from assemblages that unequivocally must include human beings and vice versa. They constitute emergent events that are always in a state of becoming other from that which they were previously. They have properties that are subject to change and alteration. They have embedded tendencies that become evident over time, and they have capacities that have been purposefully designed, as well as unknown capacities and tendencies that may be revealed over time.

Capacities are what a technology is designed to do, whereas tendencies are revealed only as the technology progresses in its actualisation. The design of a knife which takes place in its pre-individual state, for example, is designed to have the capacity to cut, in other words it must be sharp. But in use, knives have a tendency to get blunt, an emergent property we could only know through use. That discovery, however, may initiate, in the ongoing process of knife design and fabrication, a new design for a knife blade that remains sharp for longer. The use of airplanes and buses as weapons are, likewise, emergent properties beyond those for which they were initially designed. These examples serve to reveal the becoming of being, the same technology emerging or evolving into becoming something other.

## 8 Simondon and the Concept of Individuation

> The difference between the technical object and the living thing is a question of how great a latitude of becoming their past conditioning will permit. There is life when taking-form maintains itself at the brink. Life lives on a moving threshold of metastability, of fragile, provisional equilibrium that is subject to constant perturbation, from whose jaws it must repeatedly snatch its homeostasis. The living thing is an individuation that has no choice but to continue its invention, or face dissolution. Its homeostatic equilibrium is not a simple self-maintenance, but an ever-renewed achievement. (Massumi, 2012, p. 30)

Simondon was more of an influence on Deleuze than Deleuze has ever given him credit for. Deleuze wrote many books on philosophers who have influenced him but gave scant regard to Simondon. On one of the few occasions

Deleuze acknowledged the philosophy of Simondon is in a book of collected texts in which he notes that "Gilbert Simondon makes no small display of intellectual power with a profoundly original theory of individuation implying a whole philosophy" (Deleuze, 2002, p. 86).

The philosophy of Simondon is extensive and covers a number of disparate areas. However, he is also considered to be a leading philosopher of technology. This is made manifest in his book *On the Mode of Existence of Technical Objects*. However, whilst his theory of individuation is discussed in this book, he develops his theory of individuation in greater detail in his masterwork, *Individuation in Light of Notions of Form and Information* which has only in the last couple of years been translated into English.

Individuation is a theory that has the concept of becoming at its very heart. Human and technological progress are reliant upon their transformation, their inventive impulse, one that stops them from descending into some kind of digital entropy. "There is no essential human characteristic upon which an anthropology can be built [that] entails that like the human, humanism is also always a work in progress in congress with technology" (Mills, 2016, p. 3). But Simondon also argues that this relationship between human beings and technology is not simplistic. It is not enough, for example, to simply study a 'time-line' in which the invention of the bicycle is mapped out over time from then until now, with each significant change highlighted by the date it happened. Change is much more complex than this, as has been discussed above. For Simondon, it is what occurs in the liminal spaces between the dates that offer a more accurate and interesting account of 'why' these changes occurred:

> To claim to study [technologies] without an understanding of [...] the range of levels with which they interact with individuals, collectives and the wider environment seems a partial enterprise. It is as if we were to understand the Pantheon in Rome just through its symbolism and meaning without recourse to a discussion of the history of the techniques required for its building. Of course, some might argue that what is essential is the political and social structure that led to its being built. That it is cultural first and foremost. However, without the expert technical knowledge and its application to the production of the various types and density of bricks required to for its construction, the iconic dome and oculus, those wonderful symbols of the mediation between earth and the heavens, would never have been possible. Further, the infrastructure upon which that great civilisation arose and relied upon would also have been impossible. The roads, aqueducts and sewers were constructed due to technical knowledge and use of materials. This may seem obvious,

> but it is just such a blindness that infects many discussing [technology] today. (Mills, 2016, p. 4)

Here once again we find concordance with Deleuze's concept of the assemblage. Simondon likewise argues that to know something requires more than either a quantitative perspective or a qualitative perspective alone. It needs both. Its coming into being needs a much more intensive understanding of the particular milieu that is associated with its present form, together with the many milieus from the past that are likely to have influenced its coming into being.

## 9 The Relationship between Individuation and Its Associated Milieu

The coming into being, or individuation in Simondon's terms, of the bicycle as invented by Karl Drais in 1817, as an example, forms the basis for what Simondon refers to as an associated milieu. An associated milieu must never be interpreted as an environment. "The associated milieu is often wrongly understood to be synonymous with 'environment'" (Massumi, 2013, Loc 663). An environment is traditionally considered to be, in some way, a physical construct, such as an urban environment, a coastal environment or simply 'the' environment. It can also be considered in terms of an environment of fear, but again this tends to relate to a defined space within which fear is constituted. Whilst the thrust of Simondon's thinking revolves around the technical object (Simondon, 2017), an associated milieu is, nevertheless, Simondon's logic of metastable becoming, a becoming that is a regime of self-conditioning, or immanence as Deleuze would have it, or self-overcoming for Nietzsche. "This is because the regime of self-conditioning that makes technical objects [or all things] functionally autonomous [and] involves a circular causality" (Massumi, 2013, Loc 674).

## 10 The Associated Milieu

Ontogenesis, in biology, is the genesis of the individual, whereas phylogenesis is the study of groups or societies. Simondon equates ontogenesis with individuation as a process of the becoming of being. It is impossible for an individual technology to come into being without considering the milieu within which it becomes associated, and which will inevitably involve humans.

> If, [...] one presupposed that individuation does not only produce the individual, one would not seek to pass quickly through the stage of

> individuation to arrive at this final individuality which is the individual: one would seek instead to seize ontogenesis in the entire unfolding of its reality, and to know the individual through the individuation rather than the individuation starting from the individual. (Simondon, 2003, p. 24)

In her book on Simondon, Combes links the concept of individuation to the concept of the becoming of being. Combes explains that in seeking to know the individual through individuation, rather than individuation through the individual:

> The individual is thus neither the source nor the term of inquiry but merely the result of an operation of individuation. This is why the genesis of the individual remains a question for philosophy only as a moment in a becoming of being, a becoming that sweeps it along. When we retrace the genesis of physical and biological individuals or of psychic and collective reality, we always focus on the becoming of being, precisely because it is being that is individuated. As such, being can be adequately known only from its middle, by seizing it at its centre (by way of the operation of individuation and not on the basis of the term of this operation). Simondon's approach entails a substitution of ontogenesis for traditional ontology, grasping the genesis of individuals within the operation of individuation as it is unfolding. (Combes, 2013, pp. 2–3)

It is important to remember that the genesis of an individual, for Simondon, is the genesis of any-thing, not just a human. "Simondon's most basic argument is that the 'individual' is never given in advance; it must be produced, it must coagulate, or come into being, in the course of an ongoing process" (Shaviro, 2006). The 'ongoing process' means that any individuated being, natural, technological or human, is never fully realised. It will always be subject to novel, hitherto unknown and untapped possibilities; possibilities that will lead to further individuations. Referring specifically to technology, Simondon argues that individualisation is made possible:

> by the recurrence of causality [or feedback] within a milieu that the technical object creates around itself and that conditions it, just as it is conditioned by it. This simultaneously technical and natural milieu can be called an associated milieu … This milieu is not fabricated (fabriqué), or at least not fabricated in its totality; it is a certain regime of natural elements surrounding the technical being, linked to a certain regime of elements that constitute the technical being. The associated milieu mediates the

> relation between technical, fabricated elements and natural elements, at the heart of which the technical being functions. (Simondon, 2017, p. 59)

It is the feedback loop that involves what Simondon refers to as the pre-individual phase. The inspiration for this comes out of thermodynamics, where the state of something is neither stable nor unstable but, rather, sits somewhere between. This 'in-between' is what Deleuze refers to as the 'included middle.' This state is known as a metastable state. This pre-individual, metastable state contains enough potentiality that it can "produce a sudden alteration leading to a new equally metastable structure" (Simondon, 2003, p. 317; Mills, 2016, p. 36).

## 11 The Pre-Individual State

Before a technical object becomes an actual technical object, it must first be thought of, designed, invented or improved upon. These processes will involve inventors experimenting with ideas. I call this 'what if' experiments. Simondon calls it the pre-individual phase state. It should be noted, however, that the pre-individual state is not a hylomorphic state; a state where an inventor has an idea about how to form matter into something that corresponds with the inventor's idea. Indeed, Simondon goes to great lengths to dispute the concept of hylomorphism whereby "the generation of form is reducible to the imposition upon inert matter of a pregiven abstract form" (Massumi, 2012, p. 24).

Simondon's critique of hylomorphism is fundamental. Hylomorphism comes from Aristotle and consists in explaining the 'genesis' of the individual starting from the union of a matter (hyle) with a form (morphe). Sculpture, for example, may be seen as the imposition of form, from the mind of the artist, on to the marble. In the design of a temple, the observer can discern the geometry of form and granite. These observations of technical practice seem to confirm hylomorphism: the work of the artisan imposes form upon matter. Simondon argues, however, that the hylomorphic schema is insufficient when it comes to thinking of true genesis. In the case of hylomorphism, matter and form pre-exist their union; they are already of the same mode of being as the individual of which one is trying to give an account. Thus, Simondon shows that the hylomorphic schema is a conscious and an unconscious paradigm at the same time. This leads to the conclusion that hylomorphism is more social than technical. The superiority of form over matter reflects social organisation. The masters, in their observation of technicians, distinguished form from matter. Their orders contain the forms; they decide what things will become. The slaves operate on the level of matter. Like matter, they mutely await the master's orders and give them form. Hylomorphism is a mentality, a way of seeing. It is not borne out of concrete experience (Chabot, 2003, p. 78; Barthélémy, 2012, p. 212). Hylomorphism, for Simondon, operates under the illusion that that matter, be it inorganic, organic or animal, can be transformed into becoming something other, by the sheer force of human will power. Simondon (2020a) explains it as follows:

> Matter must be structured in a certain way for it to already have the properties that are the condition of form-taking, In a certain sense, we could say that matter contains the coherence of form before form-taking; yet this coherence is already a configuration with the function of form. Technical form-taking utilises prior natural form-takings that have created could be called a haecceity[5] of raw matter. A tree trunk in the timber yard consists of abstract raw matter insofar as it is considered a volume of wood to be used; only the essence to which it belongs approximates the concrete by indicating that a certain behaviour of matter will be encountered at the moment of form-taking; a pine tree trunk is not a fir tree trunk. (p. 37)

In other words, matter has a say as to whether or not it will yield to the master's imposition of form. Thus Michelangelo chose marble to create his 'David' because history told him that this was the best material, at that time, for the kind of sculpture he intended. Moreover, whilst Michelangelo prepared preparatory sketches in advance, he still would have had to yield to the natural structure of the marble, as well as being sensitive to any hidden deformations that might have emerged from within its structure. Pedagogies of transmission set within a curriculum as planned resemble a hylomorphic schema in which form (the prescribed curriculum) is imposed upon matter (all students). A nomadic pedagogy, is not hylomorphic in its delivery. There is no imposition of form on matter. A nomadic pedagogy never aspires to end up with a completed ethnotechnologically literate being. Rather, learning about technology is an ongoing, circular process, of pre-individuation phases, involving random formations of multiple territorialised forces deterritorialising and reterritorialising student voice as something other.

Simondon goes on to argue, in terms of technology, that any technical object cannot be understood independently of their 'pre-individual' phase, one that is social before actual. We cannot conceptualise actualised individuals as anything other than collections of ongoing relationships. The world is more than a sum of individuals.

> We live in a network where the pre-individual plays a significant role. Simondon speaks also of a mode of existence of technical objects. He means by this that the object is more than just any-thing. The product of an invention, is defined by its relationship with an environment that also modifies it. (Chabot, 2003, pp. 3–4)

By way of illustration, let us consider one that is not only, to say the least, extraordinary, but also one that exemplifies the tenuous link between the process of invention in the pre-individual phase, and the lack of interest leading to the unfortunate hibernation of an actualised and important technical object. The technical object in question is the development of one of the most important single mechanical devices ever invented: the crank. The crank, when combined with a connecting rod, transforms rotary motion into reciprocating motion and vice versa. The connecting rod can be a human arm or leg, as is the case with the pedals on a bicycle. Whilst various accounts of the origin of the crank abound, whether from China or Egypt, the first account of a crank in the Occident appears to be in and around A.D. 600, where it appears in the form of a rotary grindstone. What is extraordinary about this important mechanical device, was its next incarnation: the hurdy-gurdy. Lynn White Jr., takes up the story;

> A brief musical tract rather dubiously ascribed to Abbot Odo of Cluny (d. 942) describes a fretted stringed instrument which was sounded by a resined wheel turned by a crank. However, no representations of this device have been found earlier than the twelfth century, when two appear … At the beginning of the 1400's, at least twelve centuries after it was known in China and six centuries after its first appearance in Europe, the crank was still a dormant element in technology. (White, 1964, p. 110)

This most important technological development, without which the industrial revolution could not have taken place, existed but lay dormant for many centuries, other than in its use as either a rotary grindstone or as a hurdy-gurdy. Whilst the idea of the crank had been actualised concretely in these two technologies, it then became a series of fragmented memories, sleeping quietly in a virtual or pre-individual state in the background of history for almost two centuries. It was not until the early 1400s that the idea of the compound crank was awakened once more by the production of a carpenter's brace, thereby starting the process of many infinite possibilities that became actualised and will continue to be imagined and perhaps individualised in the future, "individuations determined by actual and specific differences, multitudinous influences and chance interactions" (Stagoll, 2010a, p. 75).

It is in this in-between pre-individual 'zone of becoming' (Simondon), or virtual 'plane of immanence' (Deleuze and Guattari) that we can find the relationship between the individual's ontogenesis and its milieu. Individuation is, for Simondon, an ongoing process.

> Concretisation brings not only new properties, but complimentary functions, beyond those sought after, which we might call 'superabundant functions.' [...] These properties of the object surpass expectations; it is a partial truth to say that an invention's purpose is to attain an objective, to produce an entirely predictable effect. An invention is brought into being in response to a problem, but its effects extend beyond the resolution of the problem, due to the superabundant efficacy of the created object when it is a true invention. (Simondon in Chabot, 2013, pp. 15–16)

This is why questions asked by philosophers such as Simondon, Deleuze and Deleuze and Guattari, are never deterministic or even scientific, in the classical sense. They never ask 'what is it?' The only possible answer is, for example "it is a bicycle:" a descriptive answer, an identity which is a question related to being. Rather, they ask of a bicycle 'what can become of it?' This is a question related to becoming to which the answer might variously be: a human powered flying machine (actually done already; actualised in 1962,[6] as well as in an imaginative sense in the film E.T.: a human powered and eco sustainable electric generator (done already); an eco-friendly form of transport (work in progress); a nuisance to automobile drivers (a problem to be resolved politically); a dangerous form of transport in major cities (yet another problem to be resolved politically). These are just a few of the becomings related to the bicycle that I can think of offhand.

I started this chapter by demonstrating that to know something more fully, it needs to be considered from both a quantitative and a qualitative perspective. The majority of us apparently tend, in the main, to understand the world from a more quantitatively orientated perspective. This seems certainly to be the case in craft-based technology education where the emphasis is on pedagogies of transmission. However, quantitatively orientated perspectives can lead to feelings of prejudice. For example, those who show levels of prejudice against physically or mentally disabled people, against non-white ethnicities, LGBTQ communities and children who do not conform to the expected norms, do so in the main, because they have not considered these 'other' individuals qualitatively. This absence of consideration leads inevitably to the formation of perspectives that attribute fixed and negative identities to people who do not conform to specific cultural norms. This has implications for both teaching and learning. When students are passive recipients of the knowledge conveyed by the teacher, they are disempowered. They lose the ability to act independently. Pedagogies, curriculum content and methods of assessment, like everything discussed in this chapter, are not writ in stone; they are open to becoming something other, something different, but not just different in degree, different in an immanent sense.

This chapter has considered in some detail, the concept of becoming. This will form an important philosophical element in the remainder of this book. In the next chapter I explore in more detail the human/technology relationship by asking whether *being* human is *becoming* technological or whether *being* technological is *becoming* human.

## Notes

1 For those interested in the complex mathematics of Gauss or Riemann, you can find many examples online such as https://blog.christianperone.com/2010/02/riemann-zeta-function-visualizations-with-python/

2 Virginia Woolf: The Waves.

3 When Deleuze talks about repetition, he does not mean repetition of the same, rather, he means repetition of difference. The swimmers first encounter with the water is unique to each swimmer, it is an unknown experience and so experimental. However, the next encounter will be different, and the next again and so on. Each encounter is a repetition of the one before in that it involves becoming swimmer, but with a difference. The first encounter renders the next encounter to be less alien, it is slightly more familiar and slightly more reassuring. The more the swimmer engages with the water, the more accustomed she becomes with swimming but if all her encounters to date have been in a swimming pool, for example, she may wish to swim in the sea where new experiences such as wind, water current, water temperature and waves change the dynamics of swimming. It is still repetition as she continues becoming swimmer.

4 See for example Murano Glass – The creation of the glass lion, https://www.youtube.com/watch?v=-ZNm1ZFV_OE

5 The property or quality of a thing by virtue of which it is unique or describable as 'this (one). Sometimes referred to as the 'thisiness' of something.

6 See YouTube video at https://www.youtube.com/watch?v=FQq72OvEGcw

CHAPTER 5

# Is Being Human Becoming Technological or Is Being Technological Becoming Human?

> Our bookstores are full of 'operating manuals' for every aspect of life: love, sex, divorce, friendship, raising children, eating, exercise, making money, having fun, and so on. We are our own machines.
>
> FEENBERG (2014, p. 18)

∵

## 1 Introduction

In the preceding chapter I outlined the philosophical foundations for a new nomadic pedagogy according to the work of Deleuze and Simondon. Central to both philosophies is the symbiotic relationship between humans and their technologies as outlined in the ancient Greek myth of Prometheus as told by Stiegler and discussed earlier. This relationship, over time, has resulted in many philosophical questions being raised as to whether *being* human is tantamount to *becoming* technological, or whether *being* technological is tantamount to *becoming* human.

As I have already said, ethno means race, people and culture and when conjoined to technology it means the relationship between technology and race, people and culture. I have argued that, in terms of craft-based technology education, this relationship has been, and continues to be a difficult one to realise in the technology education classroom. Because of this, I propose a new, learning *about* technology paradigm, one that seeks to study the relationship between humans and their technologies through the portal of what I call the ongoing process of becoming ethnotechnologically literate. This will require a new pedagogical structure that I will define as 'nomadic pedagogy,' a concept that I will outline later. However, prior to that, this chapter will seek to look at the way the relationship between humans and their technologies have evolved over time as an ongoing process of the becoming of being. In so doing the chapter will offer a philosophical genealogy of the ongoing relationship between humans and their technologies. In other words, this chapter is itself,

 | DOI:10.1163/9789004537002_005

an ethnotechnological exploration of this relationship, seen through the eyes of several philosophers who, over time, have elucidated very different perspectives on the subject.

## 2 Is Being Human Becoming Machine?

Across the spectrum, the concept of being human ranges from intelligent design on the one hand, to Darwinian evolution on the other. For the purposes of this book, I am, however, working on the assumption that whatever perspective is held, the one certainty common to all perspectives is that human beings came into being – somehow. Being human, therefore, starts from that premise.

The concept of 'human being as machine' is often considered to be a new, modernistic way of considering what it is to be human, usually considered under the rubric of 'post-human' or 'cyborg.' However, this same metaphor has been considered a viable way of conceptualising human being for some considerable time. The ancient Greek thinkers who preceded Plato, did not subscribe to the idea that humans were machines, although they used machine metaphors, as will be seen presently. They considered that the gods and humans were fundamentally from the same origin. In Hesiod's *Theogony*, according to (Vernant, 2006, pp. 96–97), the gods descended from the same origin as humans. "Keep it well in mind" said Hesiod (c 750–650BCE) to his brother Perses, "that gods (theoi) and men have the same origin." This has been interpreted over time, as meaning either, that gods and men belong to the same race, which is unlikely given that the gods are considered to have created humans, or that a division took place where humans may have originally shared the same existence as the gods, but became separated over time (Vernant, 2006, pp. 97–98). Empedocles (c 494-434), on the other hand, postulated, somewhat gruesomely, that human beings were not created but came about quite by chance, their various body parts wandering about aimlessly until eventually, several parts would unite in various forms:

> Here many heads sprang up without necks,
> Mere arms were wandering around without shoulders,
> And single eyes, lacking foreheads, roamed around.
> Many grew with faces and breasts on both sides,
> And man-headed bull-natured creatures, and again there arose
> Bull-headed man-natured creatures, and mixtures of male
> And female, equipped with shade giving limbs
> (Empedocles, cited in Gregory, 2008, pp. XVI–XVII)

For Empedocles, the human came about because of a multiplicity of accidents, essentially, a sort of Darwinesque evolutionary process. In stark contrast, Plato offers what might be described as the origins of intelligent design, which is also, essentially, a machine metaphor – something designed.

Plato, in his *Timaeus* defines being human as distinct and superior to all other living creatures. It was the one true god, known in Greek as the demiurge, who was considered perfect in every way, and it was he (not she?) who created the human soul, the metaphysical component of being human. The term demiurge, in Greek, signifies god as a craftsman, a creator, a fabricator. The creation of the human body, the material component of the human being which is made up of many moving parts or mechanisms, was left to the demigods to fabricate, albeit under the strict supervision of the One true god. It should be noted that the demi-gods were also created by the One true god (Plato, 2008, p. 30 (41b)). *Timaeus* offers an extraordinarily detailed account of many of the components, or mechanisms (another machine metaphor) that together, constitute the human body. In the formation of the backbone, for example, he describes how, to enable flexibility, god "lined up the vertebrae one under the other, like hinges, starting at the head and continuing through the entire trunk" (Plato, 2008, p. 74 (74a)). By using the term 'hinges,' he likens the backbone to a mechanism. Indeed, he goes further. For example, he suggests that the gods gave us legs as a form of transportation, which is why they attached legs to our bodies. They also:

> Considered the front [of the body] to be more valuable and authoritative than the back, so they made forward the main direction in which we travel ... and so, taking the head-box, they positioned the face there, fixed on the face organs to enable the soul to be fully aware, and ordained that this part our natural front, should be the leader. (Plato, 2008, p. 35 (45a))

The *Timaeus* "offers the first thoroughgoing teleological[1] account of the world, with order being imposed on chaos by an external deity" (Gregory, 2008, p. XIV). In this sense, the demiurge, or god can be thought of as a mechanic, the skilled designer of the entire universe, aided and abetted by his demi-gods, but nevertheless retaining his position as chief architect and engineer. Indeed, it is Montaigne who made the claim that it simply "not credible that this whole machine should not have on it some marks imprinted by the hand of this great architect[2] ... that the stamp of ... divinity [yet another machine metaphor] was left on the world, if only we had the wit to perceive it" (De Montaigne, 2003, p. Loc 9245; Sawday, 2007, p. 52).

Following Plato, it is Aristotle who effectively defines slaves as being inhuman lifeless things, akin to tools or machines (and it just goes on and on). There is no justice he claims, between the craftsman and his tools, nor between the soul and the body, nor the master and the slave. "The slave is a tool with a soul, while a tool a slave without one" (Aristotle, 2002 p. 158 (1161b)). Aristotle considers a slave to be 'property,' something to be owned and used to facilitate the wellbeing of the master. This pertained especially in the management of the household but included other useful functions that are more akin to the attributes of the body than to those of the mind. The slave, then, according to Aristotle, is essentially an instrument, a tool or a living animated machine to be directed or put to use:

> Now instruments are of various sorts; some are living, others lifeless; in the rudder, the pilot of a ship has a lifeless, in the look-out man, living instrument; for in the arts the servant is a kind of instrument. Thus, too, a possession is an instrument for maintaining life. And so, in the arrangement of the family, a slave is a living possession, and property a number of such instruments; and the servant is himself an instrument which takes precedence over all other instruments. For if every instrument could accomplish its own work, obeying or anticipating the will of others, like the statues of Daedalus, or the tripods of Hephaestus, which, says the poet [Homer], "of their own accord entered the assembly of the Gods;" if, in like manner, the shuttle would weave and the plectrum touch the lyre without a hand to guide them, chief workmen would not want servants, nor masters slaves. Here, however, another distinction must be drawn; the instruments commonly so called are instruments of production, whilst a possession is an instrument of action. The shuttle, for example, is not only of use; but something else is made by it, whereas of a garment or of a bed there is only the use. Further, as production and action are different in kind, and both require instruments, the instruments which they employ must likewise differ in kind. But life is action and not production, and therefore the slave is the minister of action. Again, a possession is spoken of as a part is spoken of; for the part is not only a part of something else, but wholly belongs to it; and this is also true of a possession. The master is only the master of the slave; he does not belong to him, whereas the slave is not only the slave of his master, but wholly belongs to him. Hence we see what is the nature and office of a slave; he who is by nature not his own but another's man, is by nature a slave; and he may be said to be another's man who, being a human being, is also a possession. And a possession may be defined as an instrument of action, separable from the possessor. (Aristotle, 2001, p. 1131 (1254a))

For Aristotle, then, it is accident of birth that determines the milieu in which one is destined to exist. Some, he says are marked out for subjection, others to rule. However, it is the idea that slaves are simply considered akin to instruments that suggests that one sector of society has the same qualities as a shuttle in a loom: part of a machine. George Canguilhem, in his legendary essay entitled *Machine and Organism*, highlights the fact that Aristotle drew a clear parallel between organs of animal movement [including humans] and 'organa,' or parts of war machines, like the arm of a catapult about to launch a projectile. Thus catapults, typical automatic machines of the period, seemed to be articulated like a human limb (Canguilhem, 1992, p. 48).

It is important to remember, however, in accordance with Thomas Kuhn's paradigm shift theory, that we cannot map our present-day cultural beliefs and mores onto those that existed many years ago. To adhere to Aristotle's theory of slaves as instruments, for example would today be considered horrific.

Paradigm shifts, however, occur over time and are often the result of what appear, at first, to be subtle changes in thinking. Another philosopher, this time from the Middle Ages, who described the body of both animals and humans in terms of machines, but with some important differences, was René Descartes (1596–1650).

> Examined carefully the physical economy of man: what do you find? The jaws are armed with teeth, which are no more than pincers. This stomach is nothing but a retort, or heat chamber; the veins, the arteries and indeed the entire vascular system are simply hydraulic tubes; the heart, a pump; the viscera, nothing but filters and sieves; the lungs, a pair of bellows; and what are muscles if not a system of cables and ropes. What is the ocular motor nerve, if not a pulley? And so on. Try as they will, chemists cannot explain nature and set up a separate philosophy simply by coining a new vocabulary around words like 'fusion,' 'sublimation' and 'precipitation;' for this does not at all address either the incontrovertible laws of equilibrium or the laws governing the workings of the wedge, cables, pumps as elements of mechanical theory. (Baglivi in Canguilhem, 1992, p. 47)

Although Descartes considers the human body and all animals to be machines, in his famous maxim: cogito ergo sum. (I think; therefore, I am), the only thing that Descartes believed he could know for certain, was that his mind had the capacity to think. His material body was not certain; it was possible it could be an apparition that his mind, the only thing of which he could be certain, had created. For Descartes, then, the mind, the thinking part of a human being is a separate and distinct substance, from the substance that constituted the extended body. The body, for Descartes, is an extension of the mind. Indeed, his contention that the body is a machine, which echoes Aristotle's explanation of a slave as a machine, has the effect of devaluing the body in order to justify man's using it to serve his own purpose. The mind for Descartes thus holds superiority over the body. In his *Treatise on Man* he wrote:

> These men will be composed, as we are, of a soul and a body. First, I must describe the body on its own, then the soul, again on its own; and finally I must show how these two natures would have to be joined and united in order to constitute men who resemble us. I suppose the body to be nothing but a statue or machine made of earth, which God forms with the explicit intention of making it as much as possible like us. Thus, God not only gives it externally the colours and shapes of all the parts of our bodies, but also places inside it all the parts required to make it walk, eat, breathe, enabling it to imitate all those functions which seem to proceed from matter and to depend solely on the interacting movements of our organs. We see clocks, artificial fountains, water mills and other such machines which, although only man-made, seem to move of their own accord in various ways; but I am supposing that this machine to be made in the hands of God, and so I think you may reasonably think it capable of a greater variety of movements than I could possibly imagine in it, an of exhibiting more artistry that I could possibly ascribe to it. (Descartes, 2003, p. Loc 410)

Essentially, what Descartes is implying here, is that only God can build/construct/fabricate animal machines and human beings. Other machines, other technologies, are made by man; they are 'man-made.' For Descartes, animal bodies differ from human bodies because they have no soul to drive them and no capacity for reason nor intellect. They have only rudimentary animal instincts and are, therefore, machines. Furthermore, animal machines are unable to:

> use words or other signs by composing them as we do to declare our thoughts to others … it is not conceivable that [animals] should put these words in different orders to correspond to the meaning of things said in its presence, as even the most dull-witted of men can do … it follows that it is practically impossible for there to be enough different organs in a machine to cause it to act in all of life's occurrences in the same way that our reason causes us to act. (Descartes, 2006, pp. 46–47)

Essentially, then, although using machine metaphor to describe the human body, Descartes denies that human being is a machine: that rubric is reserved only for animals. Human bodies, lacking a soul or thinking mechanism would, by argument, be animal in nature. For Descartes, then, it is the soul that differentiates human being from all other animal forms. Moreover, it suggests that God must have had some pre-existing plan (telos), some pre-existing design in mind, prior to fabricating the human machine as a body, into which he would later insert or connect its operating system; the soul or the mind.

The philosopher Gilbert Ryle in his magnum opus *The Concept of Mind*, trenchantly argues against what he calls Descartes' Myth.

> A person … lives through two collateral histories, one consisting of what happens in and to his body, the other consisting of what happens in and to his mind. The first is public, the second private. The events of the first history are events in the physical world, those in the second are events in the mental world. (Ryle, 2000, p. 13)

Ryle goes on to argue that there is a fundamental difference between the human body, which is made up of material substance and which exists in physical space, alongside other material substances, including other sentient human beings and the immaterial mind, which does not share physical space and cannot be connected in the same way as material substances. We cannot know the mind of others, other than that which they share with us. The works of dead authors and poets, therefore, induce what is known as the 'intentionality fallacy,' where living humans can only infer what the authors meant. Don Ihde extends this theme by considering the designer fallacy along similar lines (Ihde, 2014, p. 119).

Ryle introduces his concept of the 'dogma of the Ghost in the Machine' to demonstrate what he claims to be the absurdity of Descartes dualism. He claims that Dualism is a category mistake and offers several examples by way of illustration, of which the following is one:

> A foreigner watching his first game of cricket learns what are the functions of the bowlers, the batsmen, the fielders, the umpire and the scorers. He then says 'But there is no one left on the field to contribute the famous element of team-spirit. I see who does the bowling, the batting and the wicket keeping; but I do not see whose role it is to exercise esprit de corps.' Once more, it would have to be explained that he was looking for the wrong type of thing. Team-spirit is not another cricketing-operation supplementary to all of the other special tasks. It is, roughly, the keenness with which each of the special tasks is performed, and performing a task keenly is not performing two tasks. Certainly, exhibiting team-spirit is not the same thing as bowling or catching, but it is not a third

> thing such that we can say that the bowler first bowls and then exhibits team-spirit or that the fielder is at a given moment either catching or displaying esprit de corps. (Ryle, 2000, p. 18)

The rules and regulations of cricket are objective things, subject to specific definitions imposed externally. Esprit de corps is a subjective and immanent state of mind. Sometimes I would rather watch some television or go for a drink with friends than work on this book, for example. For Ryle, then, human beings are not machines.

Clearly influenced by Descartes, however, *L'Homme machine* (Man a machine) was a controversial book published in 1749. It was written by Julien de La Mettrie who, following Descartes, argues that animals are machines and, following Aristotle's Categories, which places humans in the genus 'animal,' argues that if animals are machines, as espoused by Descartes, and since humans in turn are animals, they must also be machines. However, La Mettrie differed from Descartes in that he perceived the body and the soul to be as one, independent of any transcendent controller. He went to great pains to argue for this. He also agreed with Descartes that language, and so the ability to reason, is principally what differentiates humans from other animals. He was most impressed by the French 18th century inventor Vaucanson whose ingenious mechanical duck would, when set in motion, wiggle its beak in water, quack and readjust its position. The highlight of its lifelike motion was that it could, apparently independently, eat pellets offered to it and then, after a brief moment of digestion, poop them out the other end. But he was even more impressed by the 17th century Dutch physicist, astronomer and inventor, Huyghen whose planetarium was a mechanical model of the movement of a solar system presented by modern physicists as a smooth-running machine. Vernant differentiates between machines that are designed to enable actions that humans themselves would be otherwise unable to undertake, and machines that are designed to amuse:

> Wherever the machines described by the engineers have a utilitarian purpose, they are used and conceived of as instruments for multiplying human strength, on which – despite their complexity – they depend for their only motive principle. When they call upon other sources of energy and, instead of amplifying a force given at the outset, act as automata producing their own movement, they turn out to be constructions that follow a whole tradition concerned with objects to be marvelled at, and, as such, are only marginally relevant to technology in the strict sense. (Vernant, 2006, p. 304)

It had now, apparently, become possible to take seriously the thought that "man is an automated machine that winds its own springs according to La Mettrie" (Lieber's Introduction in La Mettrie, 2016, p. 8). However, taking into account the view expressed by Vernant above, leads to the question of under which rubric man as machine would sit: that of amplifying human capability or that of designating the human as some kind of technological marvel?

La Mettrie's philosophy is, at root, deterministic in that all events are determined by previous events, but not by some transcendent creator. "These sensations which affect us decide the soul either to will or not to will, to love or to hate, these sensations according to the pleasure or the pain which they cause in us" (La Mettrie, 2016, para. 69). In other words, the human being has, for La Mettrie, no freedom of will. All actions are determined by sensations activated from external sources that are beyond the human control. "[I]t is without his permission that he enters this system where he has a place; and without his consent that, from the moment of his birth to the day of his death, he is continually modified by causes that influence his machine in spite of his will, modify his being, and alter his conduct" (La Mettrie, 2016, p. 70). The hidden mechanisms that constitute the human body are not independent of these external causes for La Mettrie. The human temperament, be it loving, hating, gentle, rough, angry, joyful and whatever passions are aroused, are for him, determined by causes that are external to the human but that affect him or her nonetheless. He concludes that: "In a world, everything should convince man that during every moment of his life he is but a passive instrument in the hands of necessity (La Mettrie, 2016, p. 72).

Although La Mettrie may now be considered a philosopher on the edges of sensibility, he was taken seriously at the time, albeit not, perhaps, because of his perceived atheism, by the church. If one were to drop the word machine from his doctrine and replace it with computer or artificial intelligence, however, it might appear to begin to have a little more credibility. Karl Popper, a renowned philosopher of science certainly thinks so. He argued that may be:

> the reason why there are so many philosophers who refuse to take the problem of physical determinism seriously and dismiss it as 'bogey.' Yet the doctrine that man is a machine was argued most forcefully in 1751, long before the theory of evolution became generally accepted, by de La Mettrie; and the theory of evolution gave the problem an even sharper edge, by suggesting there may be no clear distinction between living matter and dead matter. And, in spite of the victory of the new quantum theory, and the conversion of so many physicists to indeterminism, de

> La Mettrie's doctrine that man is a machine has perhaps more defenders than before among physicists, biologists and philosophers; especially in the form of the thesis that man is a computer. (Popper, 1979, p. 224)

Popper, on the other hand, goes on to dispute the notion of man as machine by challenging the, now famous, Turing test which he regards as an intellectual trap:

> Turing asserted that men and computers are in principle indistinguishable by their observable (behavioural) performance and challenged his opponents to specify some observable behaviour or achievement of man which a computer [a machine] would in principle be unable to achieve. But this challenge is an intellectual trap: by specifying a kind of behaviour we would lay down a specification for building a computer. Moreover, we use, and build, computers because they can do many things which we cannot do; just as I use a pen or pencil when I wish to tot up a sum I cannot do in my head. 'My pencil is more intelligent than I,' Einstein used to say. But this does not establish that he is indistinguishable from his pencil. (Popper, 1979, notes on pp. 224–225)

Bernard Stiegler also challenges the concept of dualism. For him, "The human is not a spiritual miracle that would suddenly belong to an already given body, in which the 'mental' would be grafted onto the body … [t]he psychic has its roots in a specific general physiological organisation; it is first of all a state of the body – but it is not that alone" (Stiegler, 1998, p. 144).

David Hume, the highly regarded eighteenth century Scottish Enlightenment philosopher, also compares humankind to a machine. "They [philosophers and physicians] know," he says, "that the human body is a mighty complicated machine (Hume, 2011, p. 642). He goes on to argue, moreover, that human being is only one small machinic component, situated in a machine-like universe. Another contemporaneous Enlightenment philosopher, Emmanuel Kant objects to the concept of human being as social machine. In his *Critique of Pure Reason,* he opposes the militaristic use of humankind, arguing that it is contrary to human rights:

> The hiring of men to kill or be killed seems to mean using them as mere machines and instruments in the hands of someone else (the state), which cannot easily be reconciled with the rights of man in one's own person. (Kant, 1998, p. 15)

He goes on to make the claim that to form a society of free individuals, the State must treat and consider human beings in accordance with their dignity. They must be treated as humans, not as machines:

> As nature has uncovered from under this hard shell the seed for which she most tenderly cares – the propensity and vocation to free thinking – this gradually works back upon the character of the people, who thereby gradually become capable of managing freedom; finally, it affects the principles of government, which finds it to its advantage to treat men, who are now more than machines, in accordance with their dignity. (Kant, 2016, p. 18)

However, the idea of human beings as social machines is given yet more illumination in the work of the historian and philosopher of technology, Lewis Mumford. If one assumes that the construction of an ancient city like Babylon or Sumer in Mesopotamia, or the pyramids in Egypt, would have been undertaken when the technological machinery of today was simply not available, then it is not unreasonable to further assume that the only force available to construct such large undertakings, was that involving human power. Actual technologies would have been limited to ropes, wooden rollers and limited lifting systems incorporating pulleys and levers. Except for some degree of animal power, the vast projects would undoubtedly have required a massive amount of human power, along the lines of what Mumford calls the megamachine:

> The machine I refer to was never discovered in any archaeological diggings for a simple reason: it was composed almost entirely of human parts. These parts were brought together in in a hierarchical organization under the rule of an absolute monarch whose commands, supported by a coalition of the priesthood, the armed nobility, and the bureaucracy, secured a corpselike obedience from all the components of the machine. Let us call this archetypical collective machine – the human model for all later specialized machines – the 'megamachine.' (Mumford, 2003, p. 348)

It is likely then, that slaves were used as the component parts of this megamachine. Whilst there must have been craftsmen, hence technology and technique, the largest technology at that time would have been the megamachine, which was essentially the human powered forerunner to machine technologies of the future. Deleuze and Guattari argue that this merging of the concept of human with machine is dependent upon the way it is perceived:

> The territorial machine is therefore the first form of socius, the machine of primitive inscription, the "megamachine" that covers a social field. It is not to be confused with technical machines. In its simplest, so-called manual forms, the technical machine already implies an acting, a transmitting, or even a driving element that is nonhuman, and that extends man's strength and allows for a certain disengagement from it. The social machine, in contrast, has men for its parts, even if we view them with their machines, and integrate them, internalise them in an institutional model at every stage of action, transmission, and motricity ... The same machine can be both technical and social, but only when viewed from different perspectives: for example, the clock as a technical machine for measuring uniform time, and as a social machine for reproducing canonic hours and for assuring order in the city. When Lewis Mumford coins the word "megamachine" to designate the social machine as a collective entity, he is literally correct (although he limits its application to the barbarian despotic institution). (Deleuze & Guattari, 2004, p. 155)

Canguilhem adds to the debate by citing Franz Borkenau, who conflates mechanistic perspectives within the social and economic milieus they are associated with. His focus is on the industrial age and the subsequent expansion of manufacturing:

> Once labour has been decomposed into simple, identical and easily repeatable movements, price and wages could be determined simply by comparing the hours worked – and the result was a process that, previously qualitative, had become quantifiable. Calculating work in purely quantitative terms that can be treated mathematically is claimed to be the basis and the starting point for a mechanistic conception of the life world. It is therefore by reducing all value to economic value, 'to cold hard cash,' [...] that the mechanistic view of the universe is supposed to be fundamentally a Weltanschauung [worldview] of the bourgeoisie. (Borkenau in Canguilhem, 1992, p. 50)

Moving forward in time from the 'barbarian despotic institutions' and the mechanistic conception of the life world mentioned above, to the advent of the industrial revolution and the consequential creation of more advanced and hegemonic social machines in the 'modern' era, Bauman concludes that

> The principal icon of [...] modernity [is] the Fordist factory, which reduced human activity to simple and routine and by and large predesigned

> moves, meant to be followed unquestioningly and mechanically without engaging mental faculties while keeping all spontaneity and individual initiative off limits. (Bauman, 2003, p. 101)

The idea of the social machine having 'men' for its parts, integrated into the machines and essentially becoming automated themselves, is given comic, but somewhat dystopian expression in a 1936 silent movie. It was just after the great depression that Charlie Chaplin wrote, produced, directed and starred in his ground-breaking film *Modern Times* which was described by critics as a grim contemplation on the automisation of the individual (Larcher, 2011). It depicts human beings as necessary parts of vast machine processes, in which they are merely components in an assembly production line, carrying out the same, effectively mindless, repetitive actions. Chaplin takes this even further by depicting himself becoming, literally, integrated into the internal workings of the machine. By these means, *Modern Times* offers a chilling and prescient account of how mass production, consumerism and scientific management, made famous by Taylorism, reduces human beings to mere machinic components in industrial societies.

> Taylorism was the rationalisation of labour on the model of the machine. Each movement was meticulously timed and planned to generate maximum efficiency; the system heralded a new epoch in which the worker would become as predictable, regulated and effective as the machine itself. (Murphie & Potts, 2003, p. 48)

We can see this further reflected in the famous novel *Brave New World*. In this, Aldous Huxley, describes how human beings are converted into mere cogs in a hierarchical and oppressive social machine system; one where human beings are engineered behaviourally, by means of subconscious and chemical persuasion, otherwise known as brainwashing. By these means, humans are designed to fit the needs of what is described as a perfect society, a brave new world, or utopia.[3] Technology, considered both as social and as technical machinic formations, plays a defining role in this *Brave New World*. Although it is a work of fiction, it nevertheless forms a strong correlation with what we have come to accept as the technological lifeworld we occupy today. "Once unleashed, technology becomes more and more imperialistic, taking over one domain of social life after another" (Feenberg, 2006, p. 17). Likewise, in George Orwell's dystopic novel *1984*, the populace again become social machines, programmed by, and under the control of, a totalitarian state.

From 'fiction' to actuality: in architecture, the Bauhaus movement led by Walter Gropius began to design living and working spaces in line with a minimalist and hard-edged machine aesthetic, one that rejected traditional ornamentation, thus heralding a new, modernistic style. Indeed, the French architect, Le Corbusier, coined the term for a house, "Une maison est une machine-à-habiter" (a machine for living in). Coincidentally, Le Corbusier knew and occasionally worked with the French modernist artist Fernand Léger, whose paintings were clearly industrially based, some morphing together the human and the technological. One of Léger's most iconic paintings was entitled *Les Trois femmes* (*Le Grand Déjeuner*),[4] which was bought by Alfred Barr, the first director of the Museum of Modern Art (MoMA) in New York City. In letters written by Léger in 1942 and 1943 to Barr, he describes the painting as "classic" and underlines both the universality of the subject and the clear lack of emotion. The women stare unflinchingly with blank expressions at the viewer. This, he insisted, symbolises the lack of feeling in machines (Unknown, 2020, author's translation). Citing Martinetti's '*Futurism*' poetry, Williams classifies this kind of futuristic art program as overcoming the "hostility that separates our human flesh from the metal of engines," thereby preparing mankind for "the creation of mechanical man" (Williams, 2014, p. 111).

At the turn of the nineteenth century into the twentieth, a school of thought, based on what might be described as an anti-machine discourse, evolved. This became a battle between 'pure' art and machine culture, most significantly, machine-thinking. It also sought to bring to light the cultural differences between East and West. "The machine had become a devouring monster, always attempting the destruction of handicraft" (Quennell, cited in Williams, 2014, p. 109). Part of the argument was that machine culture and fabrication tended towards the mass production of cheap products and that this led to the birth of consumerism. In a robust challenge to Quennell, on the other hand, one William McFee, a professional mechanic, proffered that "it is nonsense to speak of all things being better done by hand than by machinery. Has Mr Quennell ever ridden a hand-made bicycle, used a hand-made typewriter, played a hand-made pianoforte with hand-drawn wires?" (McFee, cited in Williams, 2014, p. 109). However, it was the potential synthesis of human and machine that was to spark a new model of being human. Esoteric Eastern culture, influenced by Buddhism, was seen to be in direct opposition to Western machine-culture. It was Ernest Francisco Fenollosa, followed by Ezra Pound and then later by craft traditionalists such as William Morris, who were to champion the Arts and Crafts Movement as part of what Fenollosa termed a new Golden Age:

> Take ordinary industry, for instance; it reasons along a thin straight line from effect to cause ... Mechanical motions follow lines of space; here moves to there, and so on forever. Numbers pile themselves up in a series that never turns around and goes backward. It is on, on, on;

> and that is why I call it a straight line. You feel the pull only one way. The place where you stop is defined by the object at the end; the value of the series is relative to a specific want. (Fenollosa, cited in Williams, 2014, p. 95)

To some extent, this perspective still holds true today. In terms of education about technology for example, this material, or purely functional machinic aspect, tends to be considered, "as mere physical objects abstracted from all relations" (Feenberg, 1999, p. 213). In this view, if knowledge development is contained or bound and presented in a hierarchical machinistic form, one that is purely neutral and functional, devoid of any creative, ethical, political or humanistic qualities other than those prescribed from upon high, then those hegemonic structures that seek to shape the technological world on our behalf, will continue to evolve and seek to guide any knowledge development in such a way that will promote a passive and compliant consumer society; a society that accepts uncritically the technological shaping of the world. I contend that this is becoming ever more the case.

Feenberg clarifies machinic Western culture by expressing his serious concerns that engineers tend to exhibit this restricted view of technology. He postulates that: "Once function is selected, the engineer has the last word on its implementation and the humanist interpreter is out of luck. This is the view of most engineers and managers; they are at home with 'function' but have no place for meaning" (1999, p. 84). The infrastructure created by engineers, from Feenberg's critical theory perspective, can, or will ultimately, be able to solve all problems by utilising machine-like principles. *Education, as I will argue later, is currently designed along these principles.* If Feenberg's concerns hold any validity, and I suggest they do, this has significant implications for epistemological and ontological development within a school subject that purports to educate about technology, whilst overtly aligning itself with engineering. This perspective reminds me of the Borg in Star Trek who assimilate other cultures into their collective machinic system. Resistance, as they say, is futile. Or is it? Fenollosa argues that although machine culture exists, humans are spiritual beings. He concludes that:

> God might have made us all machines, which cut the same undeviating groove into the metallic substance of things. But fortunately, He didn't! The very quality that distinguishes man from all things, and even from brutes, is an infinite adaptability to conditions. In war, in politics, in society, in daily life a thousand new situations arise; that is part of the fun. The problem never repeats itself. A machine can do only one thing, but

> man can do a million things, and things which he has never tried before. (Fenollasa, cited in Williams, 2014, p. 99)

So, is being human becoming technological? Certainly, the concept of humans being or becoming machines, which is technological, has thus far been simply a philosophical concept and not an actual reality. However, the question reworded as a statement becomes: 'being human is becoming technological in actuality.' This new physical conception begins its journey with human beings going beyond inventing and fabricating technologies to the invention and fabrication of technologies that are modelled upon themselves and incorporated into their bodies exosomaticaly.

## 3 Exosomatic Evolution

I start this new part of the journey by exploring the concept of exosomatic evolution. It was the biologist A. J. Lotka, who, in 1925, coined the term for the association between human and technology, which essentially constitutes a human/machine.

Techno-cultural evolution differentiates between the evolution of other animal species and that of the human being. "In place of slow adaptation of anatomical structure and physiological function in successive generations by selective survival, increased adaptation has been achieved by the incomparably more rapid development of 'artificial' aids to our native receptor-effector apparatus, in a process that might be termed exosomatic evolution" (Lotka, 1925, p. 188). Lotka clearly links human evolution with technological evolution or artificially fabricated aids that, although external to the human body, are modelled upon it. It matters not how the human came into being. It only matters, that from whatever time and by whatever processes that occurred, human beings became distinct from all other animals. This revelation came about as a direct result of exosomatic evolution.

Georgescu-Roegen went on to develop Lotka's concept of exosomatic evolution along similar lines to Stiegler's *The Fault of Epimetheus*:

> Apart from a few insignificant exceptions, all species other than man use only endosomatic instruments – as Alfred Lotka proposed to call these instruments (legs, claws, wings, etc.) which belong to the individual organism by birth. Man alone came, in time, to use a club, which does not belong to him by birth, but which extended his endosomatic arm and increased its power. At that point in time, man's evolution transcended

> the biological limits to include also (and primarily) the evolution of exosomatic instruments, i.e., of instruments produced by man but not belonging to his body. That is why man can now fly in the sky or swim under water even though his body has no wings, no fins, and no gills. (Georgescu-Roegen, 2011, p. 81)

Karl Popper, moreover, whilst still rejecting the idea that man is machine, does not reject the idea of an evolutionary exosomatic fusion of human and technology. He supports both Georgescu-Roegen and Lotka's theses thus:

> Animal evolution proceeds largely, though not exclusively, by the modification of the organs (or behaviour) or the emergence of new organs (or behaviour). Human evolution proceeds, largely, by developing new organs outside our bodies or persons: 'exosomatically,' as biologists call it, or 'extra-personally.' These new organs are tools, or weapons, or machines, or houses ... The rudimentary beginnings of this exosomatic development can of course be found amongst animals. The making of lairs, or dens, or nests, is an early achievement. I may also remind you that beavers build very ingenious dams. But man, instead of growing better eyes and ears, grows spectacles, microscopes, telescopes, telephones, and hearing aids. And instead of growing swifter and swifter legs, he grows swifter and swifter motor cars. (Popper, 1979, p. 238)

I would add to Popper's technological list of exosomatic prosthetics, another more modern example; one that enabled Oscar Pistorius, the South African athlete whose feet had been amputated as a child due to a congenital defect, to become the first amputee to compete in track events against able bodied runners at the Olympics in 2012. He gained worldwide fame for this remarkable endeavour, which earned him the nickname 'blade runner,' due to the carbon fibre prosthetic blades he wore. These blades were designed in the 1970s by the American inventor and amputee Van Phillips. Alas, Pistorius' personal life thereafter became somewhat problematic. However, that aside, it was the exosomatic evolution of the prosthetic blades, in association with his, and the designer's similar disability, that led to a co-evolutionary process; one that further enhanced the human condition. Whilst exosomatic evolution does not infer man as machine, it does offer a slight hint that points in that direction. Indeed, the human body becomes a blueprint, or perhaps more accurately, a prototype for the development of technology.

The idea of technology as prosthetic, as something artificial and external to the human body, was also the main thesis of the German Philosopher Ernst

Kapp who claimed that the "human being himself, in life and limb, represents the ideal-machinal system" (Kapp, 1877/2018, p. 137).

As early as 1877, Kapp penned what is considered by many to be his magnum opus *Grundlinien einer Philosophie der Technik* (*Elements of a Philosophy for Technology: On the Evolutionary History of Culture*). Whilst philosophical reflection on technology is as old as philosophy itself, Kapp (1877/2018) was the first to use the term 'philosophy of technology' (Franssen et al., 2018). Kapp's philosophical approach is seen very much in terms of a philosophical perspective that assumes a ready-made human. His thesis was initially predicated upon the concept of what he termed Organprojektion (organ projection), where 'organ' refers to the operational organs of the human body such as hands, feet and fingers, etc., and 'projection' refers to the technologies that enhance and develop the capabilities of the human being, via the organs with which they are associated. Essentially, Kapp differentiates between the concept of interiorisation and exteriorisation. Interiorisation represents the inner essential qualities associated with being human, particularly the operational organs, whereas exteriorisation represents the associated technological projections that exist in the outside world. Kapp explains the outside world thus:

> It is not as easy as is often presumed, not even for our sense perception, to clearly demarcate what is 'inside of us' from what is 'outside of us' … The outside world comprises myriad things that, though nature did supply the material, are less the work of nature than the work of the human hand. As artifacts, by contrast with natural products, these make up the content of the world of culture. The 'outside world' for the human being is therefore composed of both natural and man-made objects. (Kapp, 1877/2018, pp. 22–23)

Kapp argues that the outside world, although comprising many natural elements, has now become demonstrably and significantly more and more artificial: in other words, technological. And the principal organ responsible for this is, in the words of Aristotle, the human hand which he describes as the "tool of tools" (Aristotle, 2001, p. 595 (432a)). Kapp describes the hand as the human being's inborn tool, one that serves as the prototype for all mechanical tools because of its substantial involvement in the production of artifacts.[5] It is because of this that:

> A wealth of creation springs from the human artifactive drive and flows out through the hand, the arm, the teeth. The crooked finger becomes a hook, the hollow palm a bowl. In the sword, spear, oar, shovel, rake, plow,

> pitchfork, one can easily trace the dynamic tendencies of the arm, the hand, and the fingers and their adaptation to activities such as hunting, fishing, planting, and harvesting. As the stylus elongates the finger, so the lance elongates the arm, augmenting its action of force while at the same time, by decreasing the distance to the goal, also increasing the odds of reaching it – an advantage further compounded by the momentum of the lance in flight. (Kapp, 1877/2018, p. 38)

Having considered the relationship between tools and body parts, Kapp then goes on to investigate the relationship between the biology of humans and its influence on the design of machines. He uses the example of the steam engine, which he considers to be the "machine of machines" (Kapp, 1877/2018, p. 95). Whilst the steam engine cannot be analogised directly with the human, as can the simple tools discussed above, Kapp does see the machine, as an 'organ projection' in the sense of input and output. Like a living being who can convert the input energy contained in fuel in the form of food, into an output energy to be used for some animated purpose, so also, he argues, does the machine. Whilst the hand tool must combine directly with a living being, the machine is afforded much more independence by way of automation. Kapp, however, highlights important distinctions between human and machines:

> The machine is an extrinsically and deliberately made artefact, while the organism is grown ex ovo [from the egg] according to a hidden immanent law. The hegemonikon [ruling faculty of the mind; intelligence and will] of the machine does not belong to it, does not reside in it; the stoker and engine driver sit on top of it and direct it, like the rider on his horse. The ἡγεμονικόν [Greek for hegemonikon] of the living organism – intelligence and will – belongs to it, rests within it, originates with it, forms its integral constituent parts. And – without even taking its functions into account – the machine parts are there permanently; its constituent parts remain identical with themselves, until the point that external repair becomes necessary. But the organs of the organism remain self-identical in form only; their substance changes constantly, as they regenerate and repair themselves. (Kapp, 1877/2018, p. 98)

Kapp is, in effect, setting out an argument against the development of the concept of artificial intelligence, long before its inception into the digital world we now inhabit. The living being is born and becomes capable, in varying degrees, of self-organisation. In other words, the living being is a sentient being, whereas mechanisms, whether simple tools or complex machines have no consciousness: no hegemonikon. They are not self-generating and, no matter how sophisticated they may be, even today, their very existence continues to rely upon living beings, as do the uses to which they will be put. "Volition and reason, according to Kapp, are simply the highest spiritual expressions of the living being's autonomy. It is clear, not only now in which it originates, but also in particular types of behaviour that are completely unknown to machines, the most important of which is purpose-oriented behaviour" (Grigenti, 2018, p. 20).

Kapp then introduces, as his main thesis, the relationship between biology and technology as outlined in the subtitle title of his book on *Elements: On the Evolutionary History of Culture*. Kapp maintains that it is as a result of the artificially created world (technological) that culture is driven and evolves by way of "the colonisation of space (through agriculture, mining, architecture, civil engineering, etc.) and of time (through systems of communication, from language to telegraph) (Mitcham, 1994, p. 23). Kapp was greatly influenced by the potential impacts made manifest by the laying, in 1858, of the first transatlantic telegraph cable, spanning nearly 2000 miles at depths of up to more than two miles. Although this cable was later to fail, a successful cable was eventually laid in 1866. Many others were to follow around the world (History.com Editors, 2010).

Kapp's notion of a 'universal telegraphics,' considered in terms of his organ projection theory, represented by a technological analogy matching a cross section of the biological human nerve fibre system with a cross section of the submarine telegraph cable (Kapp, 1877/2018, p. 105). This invokes the idea of linking "world languages, semiotics, and inventions into a global transfiguration of the earth and a truly human habitat. But this [would only be possible] when the external colonisation of the natural environment [the transatlantic cable for example] is complimented by an inner colonisation of the human environment [Transworld communications]" (Mitcham, 1994, p. 23).

Given that technology is under constant development and ongoing re-development, Kapp argues that these associated and evolving techniques, along with the technological developments that follow, are responsible for cultural development. "It is therefore from the identity of culture and state that the power of circumstance arises. Commerce, justice, arms, police, schools, art and science, the church, posts, roads, and telegraphs are all administered by this fate" (Kapp, 1877/2018, p. 238). Essentially, Kapp was predicting the emergence of a Welt-communication (world communication) system, or what we know today as the information society, something that has had a very clear and profound impact on cultural formation and societal change. Moreover, it could be argued that the digital network we know today bears a strong resemblance to the human nervous system; essentially, what Kapp would classify as organ projection.

## 4 From Being Human to Becoming Cyborg?

In more recent times, we find a number of ambiguities emerging. Winner (1977/1992) for example, by suggesting that artifacts have politics, offers an almost anthropocentric humanisation of objects. Latour (2005) goes further by claiming that artefacts have equal agency with humans. Haraway (1991) insists that we are all becoming cyborgs. The cyborg, for Haraway, is a virtual social construction. It is neither male nor female, nor is it human or non-human: it is both male and female, human and non-human simultaneously, thus making it something 'other.' The cyborg represents a being that has not been born, nor has it been designed and fabricated. A machine (like a robot for example), is designed and fabricated but not born, whilst a human is born but not designed and fabricated (except for Plato and those who believe in intelligent design). A cyborg is both and none of these at the same time; it is thus not born, nor is it designed and fabricated. The cyborg represents the culmination of human domination over nature: the bio-technological fusion of human and machine.

Whilst, as is becoming evident, a human can conceive itself in many ways, a cyborg cannot. It is a fabrication and as such can be considered only in such terms. It has no history that is personal to itself: robots do not beget robots nor do cyborgs! Haraway contends that, in conventional terms, we can no longer distinguish between the subject (human) and object (technological or non-human). We have technologies that literally fuse into our bodies: prosthetic devices; chemicals that immunise us from disease (at least I hope so in terms of Covid 19); heart stents; xenotransplantation; genetic engineering, etc. We have technologies that attach to our bodies: spectacles; hearing aids; personal stereos; mobile phones etc. We also have technological systems that are in a state of continuous development and that serve as a means of cultural formation and reformation: transportation systems; medical technologies; the internet; etc.

The recent introduction of posthuman philosophy, however, argues that the concept of posthumanism is much more than simply a physical human-technology interface. The posthuman subject, for Hayles (1999) for example, "is an amalgam, a collection of heterogeneous components, a material-informational entity whose boundaries undergo continuous construction and reconstruction" (p. 3). Essentially, the embodied posthuman is becoming ever more infiltrated and controlled by disembodied informational systems of control. These can take the form of subliminal manoeuvring, as demonstrated in advertising projects and more sinisterly, in the manipulation of elections. Disembodied virtual networks of information form, and inform, novel networks of relationships between human and non-human entities. New, altered states of human behaviour and human cognition emerge because of ongoing and evolving networks. Our virtual posthuman world is constantly upgraded with new apps that serve to alter our habits and our habitats; shopping online (Amazon); virtual friends (Facebook) and virtual servants (Alexa).

It is information theory that drives Hayles' account which, when taken to extremes, is demonstrated in *Mind Children: The Future of Robot and Human Intelligence* by Moravec (1988). Moravec, an academic at the Robotics Institute of Carnegie Mellon University, takes the Turing test to a completely new dimension, by making the claim that it will soon be possible to download human consciousness into a computer. In other words, human consciousness could be stored as disembodied information that would never age or die but could continue to be upgraded, much like a computer. Could that be considered as being human?

Whilst Moravec's theory may seem far-fetched now, it does resonate with contemporary posthuman life in its conception of the relationship between the technologically virtual and the human. The following example suggests that being human is indeed a lesser state than being posthuman, or for Lyotard, inhuman, or for Haraway, cyborg. Ironically, we are more and more asked by our online, virtual, posthuman, cyborg, artificially intelligent entities, to confirm that we ourselves are not 'robots.' We are directed to do this by having "to demonstrate one's humanity [thereby assuming] as the central point of reference, the algorithmic culture of computational networks – not the human. This mundane example demonstrates that in contemporary society the human has become a question mark. Who or what counts as human today?" (Braidotti, 2019, p. 1). This represents a perversion of the Turing test whereby we humans are now asked to convince a set of algorithms that we are not machines.

Moravec's theory gains traction, moreover, in new and emerging theories related to the concept of the transhuman. According to Max More for example, transhumanism is the "intellectual and cultural movement that affirms the possibility and desirability of fundamentally improving the human condition through applied reason, especially by developing and making widely available technologies to eliminate ageing and to greatly enhance human intellectual, physical, and psychological capacities" (More, 2013, p. 3).

Finally, and perhaps most concerning, is the rise in the concept of artificial intelligence. This certainly causes much ambiguity around the questions of 'is *being* human *becoming* technological' and whether '*being* technological is

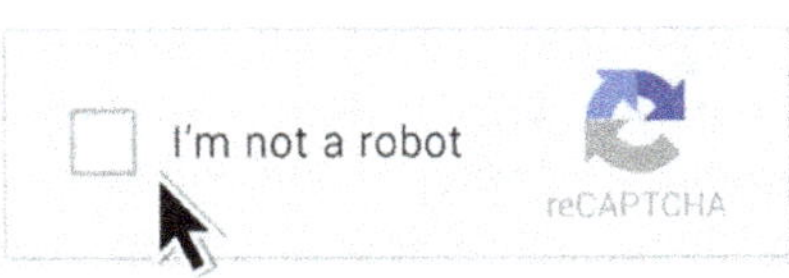


FIGURE 8
Computer tick box becoming more and more ubiquitous

*becoming* human?' An editorial in the Guardian newspaper dated Wednesday 13th March 2022, entitled 'The Guardian view of bridging human and machine learning: it's all in the game,'[6] reports on a French start-up company which may have solved AI's problem of trust with software that can learn better than humans – and expresses that as learning. In the penultimate chapter of this book, I will, however, explore Deleuze's concept of what constitutes learning and why AI does not, in his view, constitute learning.

The issues discussed in this chapter have been debated and contested for time immemorial. It becomes evident from an exploration of some of the issues involved, that the concept of being human is complex, often chaotic and certainly multistable. Without doubt, however, human beings have had, from the moment they inhabited the earth to this day, a relationship of some kind with technology. In the cave art at Lascaux in France, one image depicts a wounded bull confronting a human hunter. The bull has been injured by a spear and another spear can be seen lying on the ground beside the hunter. This spear has a carving of a bird perched on top which suggests something ritualistic, after all it has no utilitarian purpose. Clearly, these weapons are human-fabricated and are thus, artificial and technological in nature. From that time to the present day, human beings have evolved alongside their technologies, sometimes for good and sometimes for bad.

The answer to the two questions posed at the beginning of this chapter, therefore, is guided by the previous chapter on the becoming of being. The occupants of the caves at Lascaux, together with their technologies, have evolved up until the present day; that is, they have gone through an ongoing process of becoming something other up until the present, and will continue to do so into the future. Being human and being technological constitutes a transitory, multistable phase, set within an ongoing evolutionary process. Human beings invent and use technologies to cause change to occur. In that process, they disrupt the natural world, sometimes sustainably, although recently more and more unsustainably.

Being human has not significantly changed over time in terms of physical attributes; humans still have two legs, two arms and two eyes for instance. Whilst humans currently may be creeping closer to becoming cyborgs, moreover, they show no signs of mutating into becoming cyclops. Any major physical changes that have occurred in terms of being human have come about artificially and not biologically. This also holds true in terms of the natural environment, where almost every change that has come about has been artificially induced, in other words, technological. Animal habitats, on the other hand, change over time naturally through the impact of droughts or flooding for example. In modern times, however, animal habitats are also, alas, suffering serious change because of technological intervention, or more accurately, by human/technological intervention such as deforestation for example. Flooding can also be caused by technology in terms of climate change. Consequently, animals must constantly adapt to these changing environments. Humans, on the other hand rarely adapt to their environments; they adapt the environment to meet their needs, and they do so by utilising the artifice of technology. Historically, this has always happened: the only difference is a matter of degree. Today we appear to face massive problems in this respect, associated with our exploitation of the resources this planet has to offer (Dasgupta, 2021). The problem that arises, however, is that fixing things like deforestation and climate change can no longer rely exclusively on some technological intervention. Natural resources are running out which means that human reliance on technology must now be regulated, and this will require massive changes in human behaviour. Government inaction can no longer be tolerated, the dangers associated with climate change are too serious.

The issues discussed in this chapter all relate to becoming other in the ongoing process of the evolution of the human/technology relationship. As can be seen, this relationship is not at all straightforward: it is messy and chaotic. The issues discussed are more qualitative and raise very strong ethical questions. These issues are rarely, if ever, brought to light in craft-based technology education curricula. It is for that reason that I will argue that the concept of ethnotechnological literacy, which should, could and indeed can engage with the type of issues dealt with in this chapter, must be enacted sooner rather than later, but they cannot be taught in a craft-based technology education paradigm.

## Notes

1 I use the term teleological as meaning the 'end purpose' or 'raison d'être' for the existence of something. To design something teleologically, is to know in advance what its purpose or intentions are. Everything is imposed in advance nothing is left to chance.

2 Architect in Greek, is arkhitekton meaning Master Builder. Arkhi – meaning chief and tekton – meaning a builder or one who fabricates or skillful production.

3 Utopia is the title of Sir Thomas Moore's sixteenth century book. The derivation of the word Utopia comes from two contradictory Greek words meaning no place and good place, suggesting that there is no perfect world possible.

4 See the painting at MOMA online: https://www.moma.org/collection/works/79078

5 It must be here noted, that Kapp's thesis was prior to modern day artificial intelligence driven automated production systems.

6 See https://www.theguardian.com/commentisfree/2022/mar/30/the-guardian-view-on-bridging-human-and-machine-learning-its-all-in-the-game

CHAPTER 6

# Philosophical Concepts Related to Teaching about Technology

## *Introducing the Concepts of Nomadism, Transversality, the Virtual, the Fold and Subjectivity, as Crucial Aspects in Teaching and Learning the Ongoing Process of Becoming Ethnotechnologically Literate*

> Nothing has brought pedagogical theory into greater disrepute than the belief that it is identified with handing out to teachers recipes and models in teaching.
>
> DEWEY (1916, pp. 176–179)

•••

> … [N]othing can possibly be learned from an experiment that turns out just as was anticipated. It is by surprises that experience teaches all she deigns to teach us.
>
> PIERCE (1998, p. 154)

∵

My claim, thus far, is that craft-based technology education is not a suitable subject domain for the education for young people going through the ongoing process of becoming ethnotechnologically literate. This is based on the contention that the pedagogies and subject matter in craft-based technology education are incompatible with those necessary for the development of ethnotechnological literacy. Much as educators have tried, they have, thus far, been unable to coherently merge the necessary pedagogies into the delivery of craft-based technology education. I have also investigated the ubiquitous, but somewhat capricious definitions of what technology and technique actually mean to people. My conclusion is that both are context dependent and politically orientated and that any universally agreed, sustainable definition is impossible to achieve. Context and political orientation are factors in why

 | DOI:10.1163/9789004537002_006

universally agreed definitions, in almost any aspect of human life, are difficult to sustain. Some perspectives on technology, for example, consider technology as having the potential to enable everlasting life, (Morovac in the last chapter); others, in direct contrast, consider technology to have the potential to ruin the natural world (Ellul in Chapters 2 and 3).

The type of knowledge about technology that is required for the ongoing process of becoming ethnotechnologically literate, however, allows for the inclusion of the type of diverse perspectives outlined above. It falls squarely into the category that Mitcham (1994) designates as volition; a category that requires the study of social, cultural and ethical issues regarding the technologically textured world we inhabit. In this respect, Mitcham highlights the incompatibility of the pedagogies currently employed in the delivery of craft-based technology education, with those implied necessary for the delivery of what I term ethnotechnological literacy. I have in Chapter 5, moreover, indicated how the inter-relationship between humans and their technologies act symbiotically in the ongoing evolution of both, with each being dependent upon the other. Finally, I have differentiated between the concepts of being and becoming, whether human or technological. I argue that it is essential to comprehend this in order to understand the evolutionary process of how the human/technology interface has, and continues to have, a techno-cultural impact on the natural world. This is especially relevant to today and will continue to have relevance in the future; the future that the young will inherit. I make no apologies for repeating again that, faced with the existential threat to the planet, which is a direct result of the human/technology relationship, it is imperative that the young are given a voice with which to explore, discuss, debate and create ideas about how their future relationship with the technologically textured world might proceed. That voice can come about only through the adoption of a pedagogy that not only promotes but encourages a modern democratic Agora, one in which free expression is not only encouraged but actively promoted. If we continue to exploit nature for technological gain, perhaps Socrates' famous aphorism 'the unexamined life is not worth living' may, in reality, come true.

In this chapter, I will outline a nomadic pedagogy that serves to critically interrogate the human/technology relationship. I will also consider the ethico-virtuous dimensions expressed by Thales' dictum 'know thyself,' something that is, for the most part, missing in craft-based technology education at present. However, prior to that, I wish to treat this chapter as a prolegomenon to the following chapter. In so doing I will add some conceptual issues that, whilst related to the previous chapters, are supplementary, but necessary additional issues that will help link the previous chapters to the next.

## 1 Becoming Proactive and Progressive v. Being Reactive and Regressive

Two opposing modes of existence that I have previously addressed, need to be revisited at the start of this chapter. The first relates to proactive v reactive modes of existence. In the above sub-heading, I am positing that a proactive mode of existence is a vital mode of existence. It conjures up movement and change; it is becoming other. Proactive individuals take control of their lives. Synonyms for proactive include: can-do; dedicated; eager; energetic; enthusiastic; intense. Reactivity in individuals, on the other hand, denotes a lack of movement. It is much less about change. It denotes, rather, a preference for the status quo. Reactive individuals let external forces control their lives. Synonyms for reactive include; passive; inactive; retroactive.

Nomadic pedagogy rejects reactive modes of existence, favouring instead proactive modes of existence.

The second mode of existence relates directly to pedagogy. Progressive pedagogies work on the assumption that modes of existence will improve; they will get better and more advanced. They are transformative pedagogies. Regressive pedagogies, on the other hand, rely on the status quo. This implies modes of existence that are not open to change or development, or which even prefer to return to an earlier state.

Proactive and progressive modes of existence form the positive driving forces behind nomadic pedagogies. It has to be said that nomadic pedagogies, applied to the ongoing process of becoming ethnotechnologically literate, are always experimental and as such, the teacher must act in harmony with the collective forces she encounters within a class at any given time, whilst simultaneously keeping the discussions gently focussed on the subject under discussion. She must also carefully adjudicate between those opposing perspectives that can give rise to disagreement and tension between individuals or groups.

## 2 Introducing the Concept of Bildung

In his essay *Schopenhauer as Educator*, Nietzsche (2014) asks the reader "to consider … where the animal ends and where [the human] truly begins?" (p. 41). Animals, he argues, are unable to comprehend any meaning behind their lives; their lives are simply motivated by hunger and craving. They have to kill whenever they can in order to eat, and they have to flee or fight in order to survive.[1] Animal life comprises reactive forces that are, essentially, forces of unconscious, instinctive activity. Animals have the ability to communicate, but this is in a limited sense, compared to human beings. Human beings, in contrast, have developed sophisticated ways to communicate, especially by means of technology. However, any meaningful answer to Nietzsche's question of where animal ends and human begins is not so easily provided, as he himself articulates:

> The immense mobility of human beings in their vast worldly wilderness; their countless cities and states; their perenial warfare; their ceasless gathering and scattering; their hustling and bustling; their mutual imitation; their tripping people up and trampling one another down; their wailing in distress and their howling in triumph – all of it is just a continuation of animal life, as if man was deliberately reduced again and cheated out of his metaphysical endowments, as if nature, after yearning for and working towards [humans] for so long, were now shrinking from [them] and wished to return to the unconscious life of instinct. (Nietzsche, 2014. p. 42)

What Nietzsche is inferring is akin to Soctrates' unexamined life, or his famous cave analogy, both of which constitute reactive modes of existence. In this situation, people simply succumb to the will of State hierarchies by accepting the status quo, conforming to an existence that is imposed upon them externally. This leads to a reactive and unfulfilling life, one that, according to Socrates, is not worth living.

This applies equally to education. John Dewey, for example argues that "teachers should not be pedagogical slaves who passively conform to the will of authoritarian elites. He recognised that education must be practiced democratically if a society is to become authentically democratic" (Webster, 2009, p. 45). Craft-based technology education in its current format is delivered within a system designed to necessitate obedience and conformity to the ministrations of the expert teacher, the prescribed curriculum, summative forms of assessment, the regulated school and, by implication, to life beyond school. But does it have to be like this?

Stoltz (2021) elucidates what lies at the heart of Nietzsche's concern:

> Nietzsche's concern with self-cultivation (Bildung) to a large degree brings together the central tenets of his thinking to emphasise an ethics of character that is meant to serve as an alternative approach to cultivating character or the self in such a way that it reveals 'what one is' now (being), and who they could become (becoming). (p. 192)

Education, thus, involves a process of self-reflection and perspectival ways of thinking, which are both active modes of existence. In other words, it "only becomes cultivating and truly educational if it transforms our lives in such a way that we come to revalue the self as a work of art to a point where we are able to educate ourselves against our age" (Nietzsche, 2014, p. 186). What Nietzsche means by educating ourselves against our age, is probably better understood as educating ourselves against our time by understanding the successes, as well as the errors of the past. Although history helps us in this respect, Nietzsche prefers that we view history through the eyes of those great minds from the past whom we admire and have been influenced by. For Nietzsche, it was Schopenhauer. Others will have their own 'heroes' to worship. What young person who loves sport is not inspired by their hero. The same applies to music, literature and so forth, so why not for technology? It is important to note, however, that Nietzsche did not aspire to imitate Schopenhauer but chose, rather, to emulate him. This is an essentially proactive, rather than reactive, process. "The imitator [...] is without character [Bildung]; for character consists precisely in originality in the way of thinking [creativity]. He who has character derives his conduct from a source that he has opened by himself" (Nietzsche, in Golob, 2020, note 21). Or as Nietzsche (2014) goes on to say:

> How can man know himself? It is a dark mysterious business: if a hare has seven skins, a man may skin himself seventy times seven times, without being able to say, 'Now that is truly you; that is no longer your outside.' It is also an agonising, hazardous undertaking thus to dig into oneself, to climb down roughly and directly into the tunnels of one's being. How easy it is thereby to give oneself such injuries as no doctor can heal. Moreover, why should it even be necessary given that everything bears witness to our being – our friendships and animosities, our glances and handshakes, our memories and all that we forget, our books as well as our pens. For the most important inquiry, however, there is method. Let the young soul survey its own life with a view to the following question: 'What have you truly loved thus far'? What has ever uplifted your soul, what has dominated and delighted it at the same time? Assemble these revered objects in a row before you and perhaps you will reveal a law by their nature and their order: the fundamental law of your very self. Compare these objects, see how they complement, enlarge, outdo, transfigure one another; how they form a ladder on whose steps you have been climbing up to yourself so far; for your true self does not lie buried deep within you, but rather rises immeasurably high above you, or at least above what you commonly take to be your I. (p. 4)

Nietzsche offers a way for a people yet to (be)come, to discover who they really want to be, by seeing themselves reflected in an inspirational character from their past, one of their own choosing. The reflected image, however, does not connote imitation. To imitate is to copy and it is not possible to copy another individual. The reflected image should inspire one to want to equal or excel beyond the other. This is emulation and it is reserved only for a people yet to (be)come, because it will take a lifetime to achieve. Everyone knows that learning a second language is best done when one is a child; learning becomes ever more difficult as one gets older. But an important lesson to learn from Nietzsche is that to build character (Bildung), people must find the sources themselves, not have them imposed upon them by others.

Tubbs (2014) offers a wonderful summary of the concept of Bildung:

> The discipline of *Bildung* is seen as liberating, as freeing the mind from unexamined opinions and assumptions to think independently and exercise critical judgement, to question conventional doctrines and inherited claims to truth, to gain some skill in analysis and some capacity to deal with complexity, to embrace a certain scepticism in the face of dogma, and be open to many points of view. (p. 61)

Bildung, considered thus will form an important element in the nomadic pedagogy I will be proposing in the next chapter.

## 3 The Importance of Young People Having Voice

Speaking of the education of the young, in the first book of *Thus Spoke Zarathustra*, Nietzsche offers a discourse *Of the Three Metamorphoses* which outlines three different stages of the human spirit. It is important to note that Nietzsche does not invoke the Cartesian separation of the soul from the body to denote 'spirit.' Rather, what he means by 'spirit' is a mode of existence, a becoming that creates our individual character; the infinite array of forces acting upon our consciousness that give us a sense of who we are and who we might become. It is more akin to free spirit, meaning freedom from any hierarchical forces that restrict the individual from achieving what they might be capable of. Spirit for Nietzsche is what he calls the will to power which, considered another way, is individualisation through personal empowerment. This occurs as a result of continually recasting one mode of existence for another: in other words, becoming other. "In brief, spirit is a way of thinking about the living body's relationship to, and manifestation in, thought and character" (Burnham & Jesinghausen, 2010, p. 29). It is akin to Bildung in many ways.

Nietzsche's (2003) discourse outlines three stages in the process of human transformation explaining "how the spirit shall become a camel, and the camel a lion and the lion at last a child" (p. 14). I have recast the discourse somewhat, to illustrate how the human being learns about technology by entering into an ongoing process of becoming ethnotechnologically literate. This discourse does not suggest a series of stages that, if followed, will result in an end state of being ethnotechnologically literate. On the contrary, it is an ongoing process of different modes of existence, or in this case, different modes of ethnotechnological literacy.

In Nietzsche's discourse, the camel represents a beast of burden. It exemplifies 'thou shalt.' It accepts, without question, the burdens imposed upon it by the establishment. It conforms out of a sense of duty. It does not destroy, nor does it create. It is akin to a reactive mode of existence. It is not a slave in a negative sense, but, rather, it shows great strength of will and fortitude. It takes great pride in its service to the established order. It is akin to the servants of the aristocracy, particularly those in elevated positions such as butlers, who took (take) great pride in their service.[2] They knew their place in the order of things, which, by argument, infers a hierarchy, in this case of master and servant. "The weight-bearing spirit takes upon itself these heaviest things: like a camel hurrying laden into the desert, thus it hurries into its desert" (Nietzsche, 2003, p. 14). The desert signifies the loneliness of living this life of burden and compliance. For Wallen (2013), in educational terms, "[t]he modern school overflows with beasts burdened by the accretions of their lessons, the conferrals of truth, and the dutiful abeyances of public [school teachers]" (p. 197). I would argue that craft-based technology education in particular aligns itself with this perspective, especially for those who do not like craft-based technology education but have no choice but to experience it, usually in the early years of secondary school. Whilst a number of students do enjoy craft-based technology education, these are for the most part, boys. A significant number who do not enjoy the subject are, for the most part, girls. Much research, however, demonstrates that girls are more likely to take up and enjoy technology education when more emphasis is given to the social context of technology: in other words, to the development of ethnotechnological literacy (see, for example, in direct relation to technology education, Murphy, 2006; Kirk, 2014).

"But in the loneliest desert, the second metamorphosis occurs: the spirit here becomes a lion; it wants to capture freedom and be lord in its own desert" (Nietzsche, 2003, p. 14). Becoming lion is necessary because this involves fighting for its freedom: freedom to think its own thoughts; to break the chains of compliance, much akin to the student revolts in Paris in May of 1968. This is something the camel is unable to do and so it must metamorphosise into the lion. To accomplish this, however, the lion must fight the great dragon in order to break free. "The great dragon is called 'thou shalt.' The spirit of the lion, on the other hand, says 'I will!'" The dragon represents all hierarchically imposed values, regulations and rules. Borrowing from (Aoki, 2005), the dragon can be seen as representing the 'technology-curriculum-as-plan,' whereas the lion is fighting for a 'technology-curriculum-as-lived-experience.' But alas, the lion is incapable of creating new values; it is capable only of fighting the dragon and in so doing, act as the mediating force that enables the difference between being camel and becoming child to occur. This is because:

> only the child can create new values, for the child does not just eliminate what has gone before ('sacred Nay even to duty') but – by being a forgetting and an innocent new beginning – offers a 'sacred Yea-saying.'[3] The lion is a free and independent will but it is a virtue without content (so to speak), without meaning, purely negative. However, the child 'wills its own will;' the forward direction or 'virtue' of the spirit-child does not have some 'Thou shalt' as its content, but its willing is its own content. (Burnham & Jesinghausen, 2010, p. 31)

A young child has not yet been contaminated by any hierarchically imposed restrictions on its life: it is free to experience life uninhibited. How often do we give the latest and expensive 'must have' Christmas presents to our children and grandchildren, only to be bemused when, after having explored the contents of the box, they jettison the actual toy and play instead with the cardboard box it came in? The toy itself has been endowed beforehand with a specified creative content for the child to explore. Even a significant amount of Lego now comes in boxes with just the right number of bricks, together with a plan to construct only the artifacts displayed on the front cover. However, after having discarded the contents of the box, the assemblage of child/box becomes astronaut/spaceship; racing driver/formula-one car; parent/baby crib or any other creation the child/box assemblage chooses to imagine. Deleuze argues that this is what constitutes the active act of learning, rather than simply knowing learning involves creativity and imagination. Suffice to say, however, that the child rejects the toy that is infused with 'thou shalt' and 'wills its own will' that is hidden in the many folds of imagined possibilities that exist within the child/box assemblage. This represents the difference between the technology curriculum-as-plan (the must have toy with instructions to follow), and the technology-curriculum-as-lived-experience (the infinite possibilities that may emerge from the child/box assemblage) (Aoki, 2005). The child does not conform, but rather experiments and, in so doing, expresses itself differently. This is the ongoing process of becoming other, all as part of its unique evolutionary process. From my perspective and in line with the theme of this book, all of us who are involved in the education of young people, must morph from being camel into becoming lion on behalf of the young. The striated systems of governance, that continue to influence education, can prevail only for as long as we, who are involved in the education of young people at all stages, continue to act dutifully as camels. In this case, nothing will change. Our duty as educators cannot be towards policy makers: our duty must instead be to the young people we teach. This is, indeed, a perspective that Comenius held as far back as 1657:

> If anyone should wish to deliberate why God prizes children so highly, he will find no weightier reason than this, that children are simpler and more susceptible to the remedy which the mercy of God grants to the lamentable condition of men. For this reason it is that Christ commands us elders to become like little children, that is, to throw off the evil that we have gained from a bad education and from the evil examples of the world, and to return to our former condition of simplicity, gentleness, modesty, purity, and obedience. (Comenius, 1907/2010, p. 15)

Comenius does appear to foreshadow Nietzsche's argument that the spirit of the child is uncorrupted and so is open to experimentation and creative ways of thinking and acting that become corrupted only as they progress through State education and life in general. If education takes the concept of Bildung seriously, this need not be the case. It is in this respect, however, that the Nietzschean camel, unfortunately, seems to prevail in craft-based technology education (and in actuality, in all formal State education). This can be evidenced by noting how the sheer joy experienced by children in the inherently smooth space in which creative formation quite naturally takes place, begins to wither as the child progresses into primary education. This withering of creativity inexorably increases more and more through the primary stages and into the secondary sector, by which time the State has metamorphosised the smooth space of children's curiosity, imagination, creativity and experimentation, into a state-imposed striated education system in which the creative zone of experimentation gives way to the striations of State imposed conformity.

It is for this reason that I am proposing a nomadic education about technology; one that shifts the focus away from the concept of substance ontologies, that is ontologies that consider technologies independently from their affective domain, as is currently the case in craft-based technology education. Within this currently predominant framework, teaching is largely confined to issues such as, for example, the identification of a screwdriver as a tool used to insert or remove screws and a demonstration of how to use it.

I propose, instead, to employ process ontologies, ontologies that do take account of other affective domains; domains that seek to investigate the relationship between humans and their technologies. Process ontologies take account of all the forces that come together in an assemblage or milieu. It is the dynamics of the way bodies affect other bodies, whether organic, non-organic or technical. It also enables an exploration of the "possibilities of human relation, and their role in the reconstruction of subjectivity, society, and the environment. This problem circumvents the sterile dialectical conflicts between individual and society, freedom and authority, personal gain and the common good, nature and artifice" (Goodchild, 1996, p. 2). Braidotti (2013) calls this 'affirmative otherness' (p. 343). This turns these sterile dialectical conflicts into positive active forces. As Spinoza puts it; "we experience joy when a body [or the formation of an idea] encounters ours and enters into composition with it, and sadness when, on the contrary, a body or an idea threaten our own coherence" (Deleuze, 1988, p. 19). In other words, "an ethically empowering relation to others aims at increasing one's *potential,* or empowering force, and creates joyful energy in the process. The conditions that encourage such a quest involve the processes of transformation aiming at affirming positivity" (Braidotti, 2013, p. 343, original emphasis).

This approach might be considered as being characteristic of the ancient Greek Agora, a safe space where one and all could gather to discuss philosophical and political issues, or indeed any issues considered pertinent. There is a need, I suggest, for an equally important space for young people who share an interest in issues about the technologically textured world they inhabit, and will inherit, to openly discuss their concerns and feelings. I envisage, therefore, a modern version of the Agora in classroom form; one where students can safely enter debates about technology and the way it shapes their world, as well as thinking about possible sustainable technological futures. However, for this to happen, as Holland articulates in his book entitled *Nomad Citizenship*, the modern concept of an education about technology Agora, would have to be a place where

> no obvious or designated authority figure [was] standing above or amongst them [...] Whatever authority there is in the [education about technology agora] emerges horizontally. Unlike [the] Greek citizens, however, who are all alike, that is to say undifferentiated as well as equal, the citizens [in the education about technology Agora] are equal [...] yet diverse. This diversity is what distinguishes the [education about technology group] from other groups and elevates it above them.[4] (Holland, 2011, pp. 73–74)

Education, considered thus, constitutes a truly democratic education about technology, one in which vertical hierarchically imposed curricula gives way to horizontally agreed and locally devised open learning spaces where all voices have a say. This, however, is not easy to accomplish, especially in the undemocratic, hierarchically-orientated political systems that form the driving forces behind education today. Ironically, however, what is being proposed is not new. It was in 1657 that Comenius first published his magnum opus *The Great Didactic of John Amos Comenius*, in which he proffered that the "purpose of the great didactic is to seek and find a method of instruction by which teachers may teach less, but learners may learn more" (Munroe, 1900, p. 86). Comenius was proposing a more informal and democratic education that would later inspire very successful, albeit privately run, schools such as Montessori, Waldorf, Froebel, Steiner and Paideia, not to mention philosophers like John Dewey, Paulo Freire and Henry Giroux. There are also, currently, many modern examples of successful democratic schools such as Summerhill school in England, Futurum school in Sweden, the Clearwater School in Washington, the Agora school in The Netherlands and Tamariki in New Zealand to mention but a few. These schools have one thing in common: they have implemented new democratic pedagogies, not dissimilar to the pedagogy I will be proposing in the next chapter.

## 4 The Concept of Transversality

Felix Guattari is best known in philosophical circles, for his collaborations with Gilles Deleuze. However, before they met, in around 1968, Guattari was a psychoanalyst at La Bord Psychiatric Clinic in the Centre-Val de Loire region of France. He studied with Lacan but later rejected Lacanian theories of psychoanalysis. He wanted, instead, to develop a more horizontal and democratic way to interact with patients, one which was not vertical and hierarchical. He started this quest by identifying two types of groups, set within institutions. One group he calls an 'independent group.' This forms itself around an agreed collective need and controls its own behaviour. "[T]his type of group hears and is heard and can therefore work out its own hierarchising structures and so become open to a world beyond its own immediate interest" (Guattari, 2015, p. 107). In contrast, is the 'dependent group.' As one would expect, the "dependent group is not capable of getting things into this sort of perspective; the way it hierarchises structures is subject to its adaptation to other groups. One can say of the [independent] group that it makes a statement – whereas of the dependent group only that 'its cause is heard,' but no one knows where or by whom, or when [...consequently,] people [just] fit in as best they can within the situation they find themselves" (Guattari, 2015, pp. 107–112), This is much like the disaffected students, mostly girls, mentioned above. It is also apparent that, within a school setting, craft-based technology education as it is taught at present is, similarly, set within a vertical, hierarchical system with policy makers at the top and teachers near the bottom: and below teachers are the students, whom the educational edifice is supposed to be there to enlighten? As with Guattari's dependent groups, if students are heard, no one knows where or by whom, or when. They simply fit in as best they can or in the case of the dissolutioned, disrupt. The group of students that form a craft-based technology education class, then, are a dependent group, reliant upon the imposed systems that dictate curriculum content, assessment protocols and the pedagogy to be employed. These vertically orientated hierarchical systems are in direct opposition to the horizontally democratic system of transversality proposed by Guattari.

Influenced by the mathematics of transversal lines, Guattari came up with the concept of transversality. In mathematics, parallel lines never meet and so have no connection with each other. Transversal lines are lines that can, however, form a connection between parallel lines (see Figure 9).

In terms of education about technology, if the teacher is considered as being represented by line (a), and the students by line (b), as parallel lines they have no real connection with each other. There exists a disconnect between teacher and students. There is also a disconnect between the pre-ordained curriculum, which forms another parallel line, and the teacher. The teacher is obliged to teach the curriculum-as-plan, the students are required to absorb the information given. This form of teacher student relationship is described by Paulo Freire in his seminal book *Pedagogy of the Oppressed*, as the 'banking' concept of education. The curriculum content, in this model is mundane and out-of-date:

> The teacher talks about reality as if it were motionless, static, compartmentalised, and predictable. Or else he expounds on a topic completely alien to the existential experience of the students. His task is to 'fill'

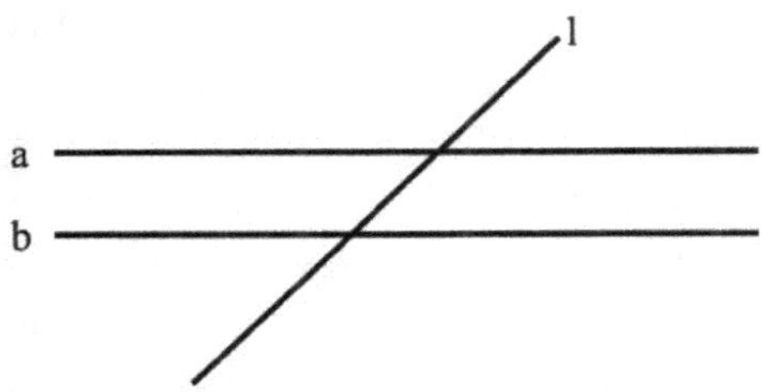


FIGURE 9
Two parallel lines a and b cut by a transversal line

> the students with the contents of his narration – contents which are detached from reality, disconnected from the totality that engendered them and could give them significance. Words are emptied of their concreteness and become a hollow, alienated and alienating verbosity. (Freire, 1996, p. 52)

The banking method is, for Freire, where content from the 'curriculum-content-as-plan' is deposited into the students' heads. Over time a withdrawal is made in the form of an examination which is designed to see if any gain in the form of interest has accrued. If, on the other hand, a connection between teacher and students is made transversally, this instigates a more democratic learning experience where the teacher and students interact democratically, whilst still retaining their own identities. More open discussions can take place, leading towards a 'curriculum-as-experience' to evolve. In this case, whilst the teacher retains control of the class, she becomes an equal co-constructor of knowledge and concepts. If other parallel lines are added and can be connected by transversal lines, new and more localised curricula can be made manifest. Lines can be made to connect department heads, head teachers, local authority education departments, initial teacher education institutes, academic researchers on education and so on.

The level of communication between these separate groups is measured by what Guattari calls the 'coefficient of transversality.' By way of example, he imagines a closed field containing a group of horses. According to research undertaken by National Geographic,[5] horses, like humans, are social animals and look after each other. When their eyes are covered, however, they tend to work independently and randomly. For the purpose of this thought experiment, I am naming each horse; policy maker; curriculum developer; assessment; head teacher; local authority education department; initial teacher education; education researcher; technology teacher. I am also adding a number of ponies that are, collectively, students. The horses are all fully blinkered to the extent that they are effectively blind to everything that is going on in the field around them. They occasionally bump into each other but for the most part, they graze, acting independently and somewhat randomly. They learn by experiencing their world alone and blind, which is clearly limiting. They construct their world but do so independently. Then along comes Felix Guattari who opens up each of the horse's blinkers, just a little. Now each horse will be able to see, to a limited extent, what is happening elsewhere in the field. They can begin to socialise a little and so, communicate just a little better than before. Now they are learning, still experientially, but because they can now partially see each other and therefore communicate much better, they are more able to learn from each other. They are now more able to socially construct their world in a cooperative and collective way. They come together as a community of learners. Imagine what might happen if we removed the blinkers altogether!

Let us now try to imagine how people act towards each other in terms of affectivity. According to Guattari, this involves more than just seeing and hearing others; it requires a democratic form of communication that reveals the way in which one group or one set of perspectives, can affect others and can, in turn be reciprocally affected.[6] This means considering issues from the perspective of a process ontology. By entering into a more horizontally orientated system of education about technology, those who held positions of power over others must modify their egotistical ideals by harnessing their super-ego.

> To accept being 'put on trial' being verbally laid bare by others, a certain type of reciprocal challenge, and humour, the abolition of hierarchical privilege and so on – all this will tend to create new group law whose 'initiating' effects will bring to light, or at least into the half-light, a number of signs that actualise transcendental aspects of madness hitherto repressed. (Guattari, 2015, p. 117)

In other words, only through a system of transversality, will it be possible to expose, and so mend, the many problems or 'madnesses' that continue to prevail in a hierarchically orientated system of education about technology: a system that gorges itself on the principle of imposition and having students act separately from each other. This propagates systems of delirium.

It is important to note, however, that the concept of transversality does not require a total surrender of power: it requires, instead, a system of empowerment for all involved. Policy makers will still make policy but now only by seriously communicating with others involved in the processes of education. This includes the students. Moreover, transversality recognises that local forms of empowerment will be different from others and must be balanced out accordingly. Schopenhauer offers a parable of the porcupines to clarify how such balance can be achieved:

> One freezing winter day, a herd of porcupines huddled together to protect themselves against the cold by their combined warmth. But their spines pricked each other so painfully that they soon drew apart again. Since the cold continued, however, they had to draw together once more, and once more they found the pricking painful. This alternate moving together and apart went on until they discovered just the right distance to preserve them from both evils. (Shopenhauer, in Guattari, 2015, p. 112)

Creating systems of equal empowerment requires the concepts of experimentation, deterritiorialisation and transversality, which when taken together, assist in the ongoing process of becoming other.

In schools, the 'coefficient of transversality' is the degree of blindness that each of the participants present, and as long as people remain fixated upon themselves, they only ever see nothing *but* themselves. This is where nomadic education can offer an alternative pedagogical framework for education about technology.

## 5 Deleuze and Guattari's Concept of Nomadism

The concept of nomadism first makes its appearance in chapter twelve of Deleuze and Guattari's book *A Thousand Plateaus* and is entitled: *1227: Treatise on Nomadology – The War Machine* (2004, pp. 387–467). In this chapter they begin by differentiating between the game of chess and the game of Go: "Chess is a game of State, or the court," they say. "Chess pieces are coded: they have an internal nature and intrinsic properties from which their movements, situations, and confrontations derive. They have qualities; a knight remains a knight, a pawn a pawn, a bishop a bishop." Craft-based technology education today follows this 'State' schema. It must, by its very nature. Learning a practical subject is akin to an apprenticeship model of learning. Indeed, the derivation of the term apprentice is from the Latin *apprehendere* and the French *apprendre,* meaning to apprehend or to learn. It is most usually associated with "training in an art, trade, or craft," (Duignan, 1998) although not always. Students learn specific craft skills associated with the fabrication of a variety of materials by using the appropriate tools, machines and methods. As has been discussed earlier, the only pedagogy that can be applied in this apprenticeship model of learning, one that is predominantly workshop-based, is that of transmission from the expert to the novice. Learning is thus restricted to the development of procedural and declarative knowledge. In addition, technology education is considered by most governments to be vocational in nature and not an academic subject. See, for example, my discussion about Dr. Peter Toft with regards to this in Chapter 4.

Deleuze and Guattari on the other hand, consider the game of Go to be less structural and conformist in nature. There are more degrees of freedom in Go:

> a Go piece can destroy an entire constellation synchronically[7]; a chess piece cannot (or can do so diachronically[8] only). Chess is indeed a war, but an institutionalised, regulated, coded war, with a front, a rear, battles.

> But what is proper to Go is war without battle lines, with neither confrontation nor retreat, without battles even: pure strategy, whereas chess is semiology.[9] ... In chess, it is a question of arranging a closed space for oneself, thus of going from one point to another, of occupying the maximum number of squares with the minimum number of pieces. In Go, it is a question of arraying [place in order or group] oneself in an open space, of holding space, of maintaining the possibility of springing up at any point: the movement is not from one point to another, but becomes perpetual, without aim or destination, without departure or arrival. The 'smooth' space of Go against the 'striated' space of chess. The nomos of Go against the State of chess, nomos against polis. (Deleuze & Guattari, 2004, p. 389)

What Deleuze and Guattari mean by nomadism, is partially related to the concept of the nomadic life of groups such as the Mongolians of the Steppes or the Sámi from northern Norway, Sweden and Finland. However, they also differentiate between the concept of nomos, polis and physis in ancient Greek culture. Holland (2011) explains it thus:

> In opposition to polis, nomos refers to space outside the city walls, a space not subject to the laws and mode of organisation of the state (or city state). It is in opposition to polis that the term nomos generates the sense of nomadism as a way of occupying space that is characteristic of nomadic peoples ... Far more often in Greek philosophy, nomos is opposed to physis. In this context, it refers to the domain of human culture as distinguished from that of nature. Thus, in a conceptual topography, the realm of nomos would lie somewhere between the unruly realm of nature or wilderness (physis), on the one hand, and the enclosed and strictly regulated space of the city or the State (polis) on the other. (pp. 14–15)

The regulated space of the State resembles the striated composition of an orchestra, whereas nomadic space articulates more with the smooth composition of an improvisational jazz group.

References to the subtitle *The War Machine* requires closer inspection. To that end, I refer to Eugene Hollands book *Nomad Citizenship*, in which he argues that there are various meanings that arise regarding the war machine. These tend to focus mainly on group dynamics and social organisation. "These include aim, object, space (smooth vs. striated), and form of sociality" (Holland, 2011, p. 25).

Amongst these variations are the rhizome,[10] a botanic metaphor given by Deleuze and Guattari in contrast to the arborescent hierarchical tree of knowledge, such as Porphyry's tree where all knowledge lies hidden in the roots awaiting discovery, whereupon it is represented in primary and general terms in the trunk, after which specialisations branch off in different directions, much like the school system. A rhizome, such as root ginger for example, has no beginning or end. It grows in ways that cannot be predicted. Break off a part and it starts to re-grow somewhere else, most likely in a different direction. Rhizomes occupy smooth nomadic space. However, the State, an example of which is traditional schooling, wants to striate smooth nomadic space by regulating and compartmentalising it.[11] The war machine in this case, serves as a metaphor to protect or rescue smooth space from the imposition of State striation (Holland, 2011, p. 25). Koch (2019) offers a fitting example of technological striation overwhelming the smooth space of nature:

> In many discourses, [there are] areas where human beings and machines are linked … into binary odd couples. Wherever there is technology, the grass of nature can no longer grow; where there is culture, we are alienated from nature; wherever nature exists, culture stops; under the sign of the age of technology arises a technocracy, the dominance of technical-calculating functional thinking, which suffocates the organically grown living world in the chilling grasp of functional administrative rationality. (p. 3)

The war machine is, thus, of a revolutionary nature, in which nomadism fights to eliminate the exploitation of the State. This involves seeking a return to a more harmonious way of living with nature. Organisations such as Greenpeace serve as examples of this. There is also a recent significant rise in protest against global warming, particularly amongst the young.

Other variations, which might be considered as having the concept of war as a central component, are in essence, the war against the State's political ends including the striating, securing and expanding of territory, as well as the exponential rise in global capitalism and consumerism. However, "Deleuze and Guattari assert at the conclusion of the 'Treatise,' and concerning precisely the supplementarity of war in relation to nomadism itself, that if nomads wage war" (Holland, 2011, p. 27), "*they do so only on the condition that they simultaneously create something else,* even if only new nonorganic social relations" (Deleuze & Guattari, 2004, p. 466, original emphasis).

> Nomadism thus has as its primary and positive object not war but rather a distinctive alternative form of organisation of social relations operating in smooth space. Hence the importance of mapping positive (creative,

> nonreactive) instances of nomadism such as nomad science [technology], and music for the project of affirmative nomadology. (Holland, 2011, p. 27)

Nomadism, from Deleuze and Guattari's perspective, is thus founded upon the concepts of becoming (Deleuze and Guattari), deterritorialisation (Deleuze and Guattari), transversality (Guattari), pre-individuation (Simondon), the rhizome (Deleuze and Guattari) and the virtual (Deleuze and Guattari).

## 6 Virtuality and the Fold

Yet another concept that informs this book is Deleuze and Guattari's philosophy of the virtual, a philosophy that will also become a vital component in the nomadic pedagogy I will propose. Deleuze and Guattari argue that, surrounding and within every actualisation, every bringing of something into being, whether social, cultural or technological, lies an essential corresponding virtual aspect which is necessary to develop a more informed understanding of the human/technology relationship. "The actual is what we experience directly. The actual is composed of identities with which we interact" (May & Semetsky, 2008, p. 145). Procedural knowledge related to the technologies and processes that make up the curriculum in craft-based technology education, means to experience directly those identifiable, actualised technologies. But, as May and Semetsky explain, "that is not *all* there is" (May & Semetsky, 2008, p. 145, original emphasis). An infinite array of *virtual* possibilities lies hidden within the folds of the actual. They are composed of an imbroglio of forces that surround, interpenetrate and compete with each other in their associated milieus or assemblages. They do this as they go through the process of becoming actual. This is a similar process to that of Simondon's pre-individuation discussed earlier.

The virtual becomes actualised by way of what Deleuze describes as the fold. Influenced by Leibniz, Deleuze explains his concept of the fold by likening it to the Japanese art of origami, or paper folding (Deleuze & Strauss, 1991, p. 231). May and Semetsky offer further clarification in understanding Deleuze's concept of the fold as well as his concept of the actual and the virtual:

> A piece of paper is folded, refolded, and unfolded in different ways to yield a particular figure. We can think of the particular figure as the actual and the paper as the virtual. The virtual is that out of which the actual comes, but it remains within the actual. The next step is to realise that while the actual is composed of identities, the virtual is not. The virtual isn't some

> kind of mirror of the actual that exists only in potentiality, so that the actual would be the virtual with the extra characteristic of its being real. The possible is opposed to the real; the process undergone by the possible is therefore a 'realisation.' By contrast, the virtual is not opposed to the real; it possesses a full reality by itself. The process it undergoes is that of actualization. (May & Semetsky, 2008, p. 145)

The concept of the fold is an important aspect of Deleuze's thinking. In French, 'the fold' translates to 'pli,' which "refers both to a twist of fabric and to the origins of life ... [and on] reflections on questions of *being* and on to the nature of *events*" (Conley, 2011, p. 192, original emphasis). The paper, which represents the virtual, is actually an imbroglio of forces, both material and immaterial, that support and compete within an assemblage (Deleuze) or associated milieu (Simondon) (see Chapter 4). These forces can be folded together in an infinite number of ways (as can the paper in origami), unfolded and refolded yet again into something other. The design of an artifact follows this concept. The forces that come together in any design result in many virtual conceptual manifestations (folds) that are considered, reworked (refolded) before ever becoming actualised materially. At this stage, one of the reworked (refolded) virtualities is eventually accepted and made manifest in actuality. It becomes an actual something. This can be in the form, for example, of a technology, an idea, a book, or a poem. Whenever the something becomes actual, there will still exist those reworkings (refoldings) that were rejected. These refolded virtualities still exist and are real, but virtual, and may form again as some other variation, or may merge to influence the design of some other product in the future, not entirely dissimilar to the case of the hurdy gurdy crank discussed earlier in Chapter 4.

For example, the invention of the bicycle, and the invention of the motor engine eventually came together in an assemblage or milieu, one which resulted in the invention of a motor bike. The bicycle and the engine were each unfolded and refolded together again, eventually becoming a motor bike. The concepts of two-wheeled transport and motor engine folded together, alongside many other forces, take place in the virtual, until one folded concept becomes accepted and actualised as one form of motor bike. The next manifestation of a motorbike will then follow and so on. This is how ideas and inventions come about for Deleuze.

This folding and refolding occurs as events in which becoming takes place. It refers also to the evolution of humans and their technologies which are in a state of constant flux, combining, decomposing and recombining to form new relationships whether positive or negative. Deleuze redefines this concept in terms of territorialisation, deterritorialisation and reterritorialisation.

Life, according Deleuze and Guattari is, as has been discussed above, not some stable, identifiable, everlasting state of being: it is a multistable state in constant flux, a series of ongoing events that affect human beings and their technologies in a multiplicity of ways. Given, therefore, that the process of evolution involves moving at various speeds from one phase to another, the becoming of being can have no original point from which to emerge. The idea that being human today is the same as being human a hundred years ago, or a thousand years ago is simply not credible. As we evolve, so too does our technological existence. Our becoming is always in conjunction with our technological development. We are not born anew into the world of our parents but are, according to Deleuze, born intermezzo: our birth constitutes a new life entering somewhere in the middle of the cosmos. Deleuze and Guattari (2004) define becoming as a line of movement:

> A line of becoming is not defined by points that it connects, or by points that that compose it; on the contrary, it passes *between* points, it comes up through the middle, it runs perpendicular to the points first perceived, transversally to the localisable relation distant or contiguous points. A point is always a point of origin. But a line of becoming has neither beginning nor end, departure or arrival, origin or destination ... A line of becoming has only a middle. The middle is not an average; it is fast motion, it is absolute speed of movement. A becoming is always in the middle; one can only get it by the middle. (p. 323, original emphasis)

## 7 The Concept of Subjectivity

Whole class teaching in technology education is teacher led, with little or no differentiation to allow for individual abilities, interests or experiences. It presupposes a form of lesson plan in which specific knowledge and skill development will be taught, learned and later assessed summatively. It is entirely predicated upon a transmission-of-knowledge pedagogy because the content of every lesson has been prepared for transmission in advance. Under these circumstances, it cannot possibly take into account the personal lived experiences brought into the classroom by each individual student. In other words, it does not recognise student subjectivities. Nor does it take account of teacher subjectivities. It only takes account of the desired learning outcomes that are prescribed for the specific lesson.

In page 1 of his book entitled *Subjectivity* (*The New Critical Idiom*), Hall (2004) begins by posing the question "Who am I?" (p. 1). In answer to this question,

I might answer 'I am John Dakers.' The concept of my subjectivity is, however, much more complex than that. I can be happy, sad, angry, in love. I can be thoughtful, forgetful, selfish, kind. Who does that make me? I am plural, I am more than *being* John Dakers, I am constantly in a process of *becoming* John Dakers.

Subjectification, for Deleuze, is a process of personal or collective events or individuations, (1995, p. 115). He also makes a distinction between individuations that are subject based: "that's me … that's you," and individuations that have no subject: a wind, an atmosphere, a time of day, a battle … One can't assume that a life, or a work of art, is individuated as a subject: quite the reverse" (Deleuze, 1995). Smith (2012) puts it thus:

> Like each of us, the philosopher – or the artist or the mathematician – begins with the multiplicities that have invented him or her as a formed subject, living in an actualised world, with an organic body, in a given political order, having learnt a certain language. But at its highest point, both writing and thinking, as activities, consist in following the abstract movement of what Deleuze calls a "line of flight" that extracts variable singularities from these multiplicities of lived experience – because they are already there, even if they have been rendered ordinary – and then makes them function as variables in order to make them function together in a singular and non-homogeneous whole, and thus participate in the construction of "new possibilities of life" – for instance, the invention of new compositions in language … and at the limit, the creation of a new world. (p. 185)

This suggests that, in a school setting for example, if a teacher identifies herself as *being* a technology education teacher, her identity is contingent upon her profession. She "IS" thus, seen by others as being a technology teacher. Given another milieu, her identity may be seen by others that she "IS" a wife, or "IS" a good swimmer. For Deleuze, "IS" is attributive to *being* something or forming an enduring identity. He asks instead that we should substitute AND for IS. She is a technology teacher AND a wife, AND a good swimmer AND … She is thus more than just a technology teacher, she is a multiplicity of events, as are we all. However, the AND does more than simply reject

> everything which could be determined as Being, One, or Whole. The AND as extra-being, inter-being. Relations might still establish themselves between their terms, or between two sets, from one to the other, but the AND gives relations another direction, and puts to flight terms and sets. (Deleuze & Parnet, 1987, p. 57)

A multiplicity does not connect attributes; it inhabits them and is inhabited by them. Multiplicities connote becoming other. Subjectivity is formed through the myriad encounters and events that affect a subject throughout his or her life. These relationships affect an individual's power to act, negatively or positively. This has clear implications for the relationship between teacher and students. In whole class transmission models of teaching, both student and teacher subjectivity is diminished. The student 'IS' a student, the teacher 'IS' a teacher. Nomadic education, on the other hand, regards the fragility of participant subjectivity as being metastable and not a fixed state of being. For nomadic education to be successful, teachers must become self-reflexive and so ethical in their associations with students. These ethical concerns must permeate the teachers' strategies because, as Semestky and St Pierre argue, "the pluralism of significations breaks down any *a priori* definition of ethics as transcendental and being the same for everyone in every situation. Rather, ethics explodes anew in every circumstance ... The self is not given ... there is no core, essential self that remains the same throughout time ... subjectivity is constructed within relations that are situated within local discourse and cultural practice" (St Pierre, in Semetski, 2006, p. 96).

## 8 Pedagogies for Learning Ethnotechnological Literacy

The transmission model in craft-based technology education is by its very nature a monologue, where interaction between teacher and pupil is a one-way process. There is a growing recognition, however, that for children to learn, they have to be actively involved in the learning process. They construct meaning through the process of interaction and inquiry, which involves communicative action. "We have been moving away from cognitive theories that emphasise individual thinkers and their isolated minds to theories that emphasise the social nature of cognition and meaning" (Resnick, in Barab & Duffy, 2000, p. 26). Learning then, does not take place in a vacuum. Children do not learn simply by constructing their own individual realities in isolation from the cultural, historical, and social environment into which they were born. Without those factors there is no conceptual framework to work from (Resnick, in Barab & Duffy 2000).

Resnick (in Barab & Duffy, 2000) compares learning in schools to the acquisition of knowledge metaphor and makes an important distinction between this and the spontaneous and tacit learning that occurs outside school. She found that the contextualised, concrete and in particular, collaborative form of learning that takes place outside school, engendered a much richer learning experience than that encountered inside the school, where learning tended to be prescribed, largely decontextualised and as a result, abstract in nature. Children moreover are not merely learning machines that soak up information until they reach adulthood; they also contribute to the creation of knowledge as part of the process.

Learning is, fundamentally, constituted through interactions and relationships in a given sociocultural milieu (Cole, 1996; Engeström et al., 1999; Lave, 1993; Lemke, 1997; Matusov, 1998; Rogoff, 1990; Vygotsky, 1978; Walkerdine, 1997; Wenger, 1998; Wertsch, 1990). This milieu comprises a variety of particular cultural identities situated in a particular environment, whether natural, social or artifactual, in which a community of practice, and thus learning, is constituted and where "(p)ractice is not conceived of as independent of learning" (Barab & Duffy, 2000, p. 26).

Mapping these ideas onto those of Semetsky, the consequences of introducing these perspectives into the teaching about technology curriculum, "proposes a meaningful pedagogy of values, with values *per se* continuously revaluated and created anew in the manner of signs as a function of situations, events, and diverse experiential and experimental contexts" (Semetsky, 2017, p. 9). Or as Deleuze puts it in relation to understanding why contrasting values exist, we need to evaluate the value of values. This can only ever be achieved by communicating with those who have contrasting perspectives and seeking out the origin of the mode of existence that created them in the first place (Deleuze, 2006, pp. 1–2). For example, consider a lesson operating under the auspices of a nomadic pedagogy discussing climate change. Several students suggest that we should ban the use of SUVs immediately and permanently. Several others object forcefully suggesting that would act against their human rights and freedoms. Left at that, students would most likely continue to hold those opposing perspectives. However, by initiating what Nietzsche calls perspectivism, whereby each individual recognises and accepts that the world is full of differing and opposing perspectives, discussions could then take place to find out the genesis of each-others perspective. Whilst this may not necessarily lead to any resolution, it may lead to the discovery of some aspects in each-others perspectives that overlap, thus leading to some degree of rapprochement occuring. Considered thus, ethnotechnological literacy involves the need for young people to be given the opportunity to explore their personal values by expressing their thoughts, concerns and perspectives and by thinking of possible alternative futures through the process of folding and unfolding. After all it is they who will inherit the future. Ethnotechnological literacy, in this form, is the ability to anticipate and create possible futures – through improvisation, experimentation, imagination and invention. It can also provide a democratic space, an Agora that encourages reasoned, conciliatory, and constructive ways to express dissent against extant systems that threaten the(ir) future, such as the crisis of global warming. In this case, such a forum will also hopefully enable them to consider ways that sustainable futures might be made manifest.

What follows are two more detailed examples of scenarios that could be discussed under the auspices of nomadic pedagogy relating to learning about technology. Both lend themselves to the ongoing process of becoming ethno-technologically literate: one historical and one fictitious.

## 9 Moby Dick and the Lorax

In the late 19th century, whaling became what can only be described as industrialised. Prior to this, whaling had been conducted for a long time, but on a sustainable level. The principal reason that whaling grew to become such a huge and important industry was because of the demand for whale oil which was used to light lanterns and gas lamps during the transition from candle power to the development of electricity. Whaling became so intense at that stage that certain species of great whales are to this day listed as endangered and are at risk of extinction unless we humans intervene with some sustainable solutions. This is just one example of the commercial exploitation of natural resources made possible by the advancement of technology in the service of efficiency and capitalism. Industrialised whaling used larger ships and specialised technological equipment for harvesting as great a number of whales as possible, thus paving the way for huge profits to be realised at the expense of the natural world. This was further aided by the availability of cheap labour (Unknown, 2008). However, although this industrial scale exploitation of nature enabled profit to mutate into capitalism, then capitalism into neoliberalism, the whaling profiteers of those times had no idea of the impact their actions were having on the natural world. However, we do know now, and this applies to many of the situations in which we find ourselves today. How might we resolve this?

Fifty years ago, Theodor Geisel (aka Dr. Seuss) wrote a book for 'children' entitled *The Lorax* (Geisel, 1971/1999). It was turned into an animated movie in 2012. "It is set in a dark, post-apocalyptic landscape, destroyed by [the] excessive exploitation of natural resources [and] makes connections between capitalism, greed, and environmental degradation" (Axell, 2018, p. 899). It is interesting to note that the 50th anniversary of the publication of this book was the year 2021, the same year that the United Nations COP26 conference was held in Glasgow. Although not many references were made to the book, certainly not from our world leaders, I found two. One was from the Washington Post.

This reported that "Nori Nicholson, 7, from near Edinburgh, and her brother Ted, 5, walked with homemade Dr. Seuss-themed signs. On one of them, the Lorax says, "I speak for trees." "It's their future we are talking about here," their mother, Amy, said when asked why they made the trip. "We can't be silent" (Dennis et al, 2021). The second was from a speech given by Adi Mariana Waqa from Suva, Fiji on the 6th November 2021 at the COP26 march for creation. She closed with a quote from the Lorax aimed at the world leaders attending the conference "Unless someone like you cares a whole awful lot, nothing is going to get better. It's not" (Geisel, 1971/1999, back cover). The whole book, although written half a century ago, presaged the world as we find it today. As Blair indicates:

> "He (Dr. Seuss) wanted a book that captured the effects of pollution on ecosystems and I would say it was really ahead of its time," says anthropologist and evolutionary biologist Nathaniel Dominy who teaches at Dartmouth. "The different species disappear from the narrative in succession," he notes. "The Bar-ba-loots leave because they run out of food. The Swomee-Swans leave because the air is polluted. The humming fish leave because the water's polluted. He's describing what we would now call a 'trophic cascade' and for me, as a scientist, I just find that genius that he anticipated that concept by a decade or more." (Blair, 2021)

Each of these scenarios lend themselves to providing material to introduce the ongoing process of becoming ethnotechnologically literate. The Dr. Seuss book is directed towards children in primary schools but, I would argue it also has great relevance for secondary students. The film of the book is pretty good too. A consideration of the impact of whaling would provide an interesting introduction to a consideration of the impact of technologies on the environment for older students.

Following on from the previous chapters, this chapter has brought together some concepts that will serve to reinforce those already discussed, particularly in relation to the concept of becoming. Concepts of proactivity, progressiveness, nomadism, transversality, the virtual, the fold and subjectivity are all issues that support a nomadic pedagogy. This chapter serves to enhance the arguments made in previous chapters regarding the concept of becoming as an important feature for the development of the ongoing process of becoming ethnotechnologically literate, whilst simultaneously, acting as a prolegomenon to the next chapter on nomadic pedagogy.

## Notes

1 I am referring to animals in the wild here. Obviously domestic animals and farmed animals do not altogether live the same life as wild animals. Nor do those in zoos where food is supplied by humans.
2 The British/American television production of Downton Abbey serves as an interesting example.
3 A nay sayer opposes, denies, or takes a pessimistic or negative view, usually opposing something or saying no to it. A Yay sayer, on the other hand, has a confident affirmation, a positive attitude who tends to accept uncritically the perspective of others. Any criticality is dealt with in a positive fashion.
4 In this citation, Holland is discussing progressive neighbourhood activism. I have simply restructured it to focus on education about technology.
5 See https://www.nationalgeographic.com/animals/article/horse-communication-ears
6 Affect is usually a verb, and it means to impact or change. People are affected in different ways by technology the effect of which may be good or bad. Effect is usually a noun, an effect is the result of a change.
7 Synchronicity is a concept introduced by Carl Jung to describe circumstances that appear meaningfully related yet lack a causal connection (Wikipedia). In Go it is extremely difficult, if not impossible to understand the cause that enabled the winner to succeed. Indeed, it is only very recently that an algorithm was developed that allowed a computer to play Go and even win on occasion. Go is pure strategy.
8 Diachronic is contrasted with synchronic in that it is possible in chess, for example, to follow and comprehend the game as it develops and evolves over time. Chess masters learn by studying and remembering moves in other games. Chess is semiology.
9 "We can define semiology or semiotics as the study of signs. We may not realize it, but in fact semiology can be applied to all sorts of human endeavours, including cinema, theatre, dance, architecture, painting, politics, medicine, history, and religion. That is, we use a variety of gestures (signs) in everyday life to convey messages to people around us, e.g., rubbing our thumb and forefinger together to signify money" (Seller, 2021). "For example, the menu we consult in a restaurant has been drawn up with reference to a structure, but this structure can be filled differently, according to time and place, e.g., breakfast or dinner" (Barthes, 1964, p. 28).
10 The concept of the Rhizome is given in the first chapter of Deleuze and Guattari's book *A Thousand Plateaus: Capitalism and Schizophrenia* (pp. 3–28).
11 School subjects are compartmentalised and rarely communicate with each other, especially from the student's perspective. For example, design is taught in design and technology as well as in art but there is rarely communication between the departments.

CHAPTER 7

# What Are the Characteristics of a Nomadic Pedagogy for Teaching about Technology?

In the preceding chapters I have highlighted Deleuze's assertion that thinking is about the creation of new concepts and not just a repetition of pre-existing ones. In terms of craft-based technology education, repetition consists of learning about what existing technologies are and what they do (tools, machines, electronic components, computers, cut, saw, shape, CNC, etc.), as well as the development of skills in relation to the use of such tools in the fabrication, and in some cases the design, of low-tech technological artifacts. The ongoing process of becoming ethnotechnologically literate, in contrast, has as its *raison d'etre*, the encouragement of young people to think for themselves. School students, according to Lipman et al. (1980), "frequently complain that their courses lack relevance, interest, and meaning" (p. 5). A significant proportion of students, particularly girls, feel this way about craft-based technology education (Dakers et al., 2009; De Vries, 1986; Dow, 2006; Murphy, 2007; O'Riley, 2003; Paetcher, 2007). Lipman et al. (1980) then go on to make a clear and important distinction between what might be termed thinking, and thinking for oneself.

> We can study the processes of thinking in a detached, impersonal, and objective fashion ... but it is quite another matter to reflect upon and bring to utterance our own personal perspectives. In this sense, thinking for oneself involves reflection upon one's own experience and upon one's own situation in the world. It requires appraisal of one's own values and in effect of one's own identity. It further involves a search for more and more reliable criteria so that the judgements one makes in the course on one's life can rest upon a firm and solid foundation. (p. 203)

Mathew Lipman and Robert Fisher were pioneers in teaching children to think by introducing methods of philosophical enquiry into the classroom (Lipman et al., 1980; Fisher, 2000). None of these methods, however, has ever featured in technology education. Even earlier, John Dewey (1916) postulated that "all which the school can or need do for pupils, so far as their minds are concerned ... is to develop their ability to think" (p. 191). Thinking for oneself can be considered to be an unfolding of the world to reveal alternatives, or to

 | DOI:10.1163/9789004537002_007

fortify individually held truths. However, according to Ronald Bogue (2008), a renowned Deleuzian scholar, "the search for truth always commences with a disruptive event that compels thought into action" (p. 2). An example of an intense disruptive event is that of the Apollo 13 mission. NASA had a great deal of data and experience in sending a crew to the moon and back again. However, they had no data about what to do in the case of the events that unfolded after the explosion. Consequently, they had to come up with a variety of solutions to a multitude of problems in lived time as the plethora of problems unfolded. This compelled individual scientists and engineers to express their own thinking as part of an active collaborative process of real-life problem solving.

This example also highlights the fact that thinking *for* oneself is not the same as thinking *by* one's self. The technologically textured world has come about only by collaborative processes of innovation, some good, some bad. This is a problem for craft-based technology education in which each student learns, works and is assessed independently. There is little if any collaborative endeavour. A nomadic pedagogy, in contrast, can operate successfully only by the formation of a community of inquiry and this necessitates collective participation.

> Participating in a community of inquiry engages young people in important cognitive moves such as creating hypotheses, clarifying their terms, asking for and giving good reasons, offering examples and counterexamples, questioning each other's assumptions, drawing inferences, and following the inquiry where it leads. But inquiry is also a social enterprise, which requires students to share their own perspectives, listen to one another, read faces, challenge and build on one another's thinking, look for missing perspectives and reconstruct their own ideas. This kind of meaningful classroom dialogue is something most students find irresistible: they can't help joining in, contributing their own reflections. In this way, cognitive and social skillfulness are acquired naturally and in context, rather than in isolated drills. (I.A.P.C., 2022)

This chapter seeks to ask the question, "what are the characteristics of a nomadic pedagogy for teaching about technology?" I have already discussed Deleuze and Guattari's concept of Nomadism in the last chapter. I will later in this chapter discuss how the concept of a nomadic pedagogy is associated with the ongoing process of becoming ethnotechnologically literate. However, prior to that, I want to briefly discuss the concept of pedagogy in more general terms. This is a term that everybody knows, especially those involved with education, whilst at the same time, it is a term that few people are able to articulate meaningfully, especially those involved with education. Van Manen offers the derivation of the term thus:

> "Agogic, -agogue, agogy," are rare Greek root words that mean or connote "guiding, eliciting, drawing forth, accompanying, helping, assisting, leading." Agogy, as a social science, studies how people change; agogical practices provide directions for interventions in helping people transform themselves and develop self-responsibility and mature independence. Pedagogy is the agogy of children (paides means child); ortho-pedagogy is caring for children who have special problems or handicaps (such as in the context of special education). Andragogy is the agogy of adults (andra means older). For example, university or other adult-oriented educational programs have an andragogical focus. We can also speak of agogical care when concerned with caring for the elderly, such as in nursing homes. An interesting question is, of course: what forms of knowledge, qualities, and ethical values are required for someone involved in agogical practice? (van Manen, 2015, p. 41)

I will specifically address the answer to van Manen's last question posed, when I discuss my version of nomadic pedagogy as employed in the ongoing process of becoming ethnotechnologically literate.

Van Manen goes on to differentiate between instruction, that is teaching the curriculum as planned and pedagogy which involves a curriculum as lived experience (Aoki, 2005). The latter "has no external goal outside of itself" (van Manen, 2015, p. 42). Educational goals are promulgated by external agencies such as policy makers, as I have already discussed. In contrast, pedagogy is the actual activity itself, one that acts in the best interest of the learner. Unfortunately, craft-based technology education is locked into a curriculum as planned, which, as van Manen (2015) stresses, means that when "pedagogical action [...] is motivated by the external setting of learning outcomes and achievement goals [it] inevitably turns into instrumental action, action in the service of calculative ends." This is in contradistinction to (ped) agogy of young people, as mentioned above, as a process of helping people to transform themselves and develop self-responsibility and mature independence. In terms of pedagogy, "We not only teach students knowledge and skills; more importantly, perhaps, we teach them to recognise who they are and what they are. We teach them about themselves, to reflect on their own selves, and how we regard them" (van Manen, 2015, p. 121). Pedagogy is, thus, difficult for teachers to achieve, especially in this time of summative assessment and the inspection of schools, the outcomes of which are predicated upon the success of teachers getting students high grades in their exams. Even Heidegger (1972) is aware of the difficulties:

> Teaching is even more difficult than learning. We know that; but we rarely think about it. And why is teaching more difficult than learning? Not because the teacher must have a larger store of information, and have it always ready. Teaching is more difficult than learning because what teaching calls for is this: to let learn. The real teacher, in fact, lets nothing else be learned than – learning. His conduct, therefore, often produces the impression that we suddenly understand merely the procurement of useful information. The teacher is ahead of his apprentices in this alone, that he has still far more to learn than they – he has to learn to let them learn. The teacher must be capable of being more teachable than the apprentices. The teacher is far less assured of his ground than those who learn are of theirs. If the relation between the teacher and the taught is genuine, therefore, there is never a place in it for the authority of the know-it-all or the authoritative sway of the official. It still is an exalted matter, then, to become a teacher .... That nobody wants any longer to become a teacher today, when all things are downgraded and graded from below (for instance, from business), is presumably because the matter is exalted, because of its altitude. And presumably this disinclination is linked to that most thought-provoking matter which gives us to think. We must keep our eyes fixed firmly on the true relation between teacher and taught – if indeed learning is to arise. (pp. 15–16)

Note that for Heidegger, if indeed learning is to arise, it cannot be dependent upon lesson plans or learning outcomes. It relies, instead, upon the relationship between the teacher and the students. For Dewey, that relationship must be, in addition, democratic. A democratic pedagogy, in terms of learning, takes place between adults and young people, where all participants are considered to be learners. The adults involved in pedagogy are not confined to teachers alone, they involve all adults that enter into a mutual learning experience with young people. However, it is the adults who are responsible for guiding, eliciting, drawing forth, accompanying, helping, assisting and leading the learning process. In any learning process so conceived, students

> must be able to involve themselves fully, openly, and without bias in new experiences. They must be able to reflect on and observe their experiences from many perspectives [Nietzsche's perspectivism]. They must be able to create concepts that integrate their observations into logically

> sound theories, and they must be able to use these theories to make decisions and solve problems. (Kolb, 1984, p. 30)

Nomadic pedagogy, as I will propose momentarily, operates as discussed above. Pedagogy in this sense, understandably, solicits fear and trepidation from policy makers and teachers alike. The former, seem to believe teaching to the test is just fine as it is, whilst teachers continually find themselves suffocated by these same extant out of date practices. Another concern shown by teachers, is the idea of 'losing control' by giving students more degrees of freedom. This is perceived to be anathematic to student discipline. Van Manen explains this complexity thus:

> Many pedagogies are dealing with the issue whether one should care for the child through arrangements of dependency or let the child be independent. From the point of view of pedagogical antinomies,[1] this conflict between dependence versus independence is a false dichotomy, and a construct of binary thinking. Children who are kept totally dependent do not thrive, but children who are given total freedom do not thrive either. Independency and dependency, freedom and control, constantly must be brought into a certain harmony, relative to the child's age and maturity. Children can only become independent through dependence. The question therefore is not dependence or independence but rather what kind of dependence the child needs in certain situations for the sake of growing independent. No human person is probably totally independent; even adults maintain relations of dependence with their life partner, lover, relatives, friends, and communities. In our experience these antinomies are intermingled and not really separable. (van Manen, 2015, p. 125)

Pedagogy, considered thus, is about transformation and change. It is an immanent form of becoming other, a process that is not imposed upon a learner, but involves learners in the context of their own experiences and those new experiences that they might encounter in the future, whether in the classroom or out.

I have only touched upon the complexity of pedagogical theory relating to practice. I have given this brief account to show that philosophical thinking in this respect, is much in alignment with my version of nomadic pedagogy.

I now want to look at the characteristics of technology education as espoused some 36 years ago in Eindhoven. In 1986, the very first PATT (Pupils' Attitudes Towards Technology) conference was held in an attempt to discover what the characteristics of technology education were, and what girls and boys thought about technology. In his presentation, Marc de Vries indicated that the subject should entail five characteristics that should have a place in the curriculum. The conclusions were that:

1. The pupils should have to learn to give value-judgement to technology from their own view of life. They must learn to see that technology is not a specific operating field for men.
2. In education, attention must be paid to the historical aspect of technology ... it is mankind, not the machine, that is at the centre of technology. Technology has to be presented in all its diversity.
3. The relationship between [science] and technology has to be made clear.
4. The pupils should learn three skills: designing, practical technical skills and handling technical products.
5. Pupils must be impelled to think about the positions of technology in society. Their own responsibility in the association with technology and the fact that pupils are also consumers, are part of this as well (De Vries, 1986, p. 17).

Craft-based technology education has, for the most part, aligned itself with the characteristics of item 4 above. None of the other characteristics is evident, in classroom practice nor in curriculum content. This has been discussed in previous chapters. As far as this book goes, in terms of the ongoing process of becoming ethnotechnolocally literate, I will happily adopt the characteristics proposed above, with two exceptions: First, that characteristic number two be changed to:

2. In education, attention must be paid to the historical aspect of technology ... It is both the historical and ongoing relationship between humans and their technologies that is at the centre of technology. Technology has to be presented in all its diversity.

Second, that characteristic number 4 be left to the purview of craft-based technology education. My reasons for this been discussed in previous chapters. I would further add one more characteristic I deem to be important in teaching about technology: that is that students need to think about how politics effects and affects them in their technological lifeworld both of today and of the future. Politics plays an important role in the way technology impacts upon economies, sometimes for good, sometimes for bad. Global warming serves as an example of the latter. This resonates with de Vries' number 5 and is supported today by UNESCO's (2020) call for the inclusion of ESD (education for sustainable development) within all school curricula as follows:

> The technological future: ESD has to respond to the opportunities and challenges brought about by technological advances. Some 'old' problems will be resolved through technology, but new challenges and risks

> will arise. Critical thinking and sustainability values become ever more relevant, as the task of teaching ESD may become more challenging with the illusion that technologies can resolve the majority of sustainability problems. (p. 18)
>
> Education policy-makers should embed ESD in education quality assurance criteria in order to ensure that education institutions are monitored and assessed for progress on how well they develop learners' capacities as change agents. (p. 26)
>
> Policy makers from the sustainable development sectors should integrate ESD into all policies that explicitly address the achievement of the SDGs [sustainable development goals], in collaboration with education policy-makers, civil society organizations, private companies, and academia. For example, ESD should be prominently placed in all policies that address climate change (SDG13). These efforts will require close collaboration between policy-makers from the sustainable development and education sectors, in particular among relevant ministries, as well as strengthened collaborative governance among diverse stakeholders. (p. 26)

The concept of a nomadic pedagogy that enables the ongoing process of becoming ethnotechnologically literate is structured to exactly meet the UNESCO ESD requirements stated above, along with many more technologically orientated issues.

It is clear that now, more than ever, we need to engage young people and teachers in the process of learning *about* technology by utilising a new and progressive nomadic pedagogy, and whilst I have outlined the characteristics of teaching about technology, what now follows is a consideration of how these characteristics can be achieved.

## 1 A Nomadic Pedagogy of Education (Bildung) about Technology

> The world will never starve for want of wonders; but only of want of wonder. (Chesterton, 2007, p. 10)

### 1.1 *Introduction*

One of the problems with current university and school education in general, and with technology education in particular, is that both the curriculum and pedagogy are being driven by business models that align teaching and learning to the workplace (Bonta, 2013; Deleuze, 2004a; St Pierre, 2008). Deleuze presaged this in the 60s and 70s. "One can envisage education becoming less and less a closed site differentiated from the workspace as another closed site, both disappearing and giving way to frightful continual training, to continual monitoring of worker-schoolkids or bureaucrat-students. They [the State] try to present this as reform of the school system, but it is really dismantling" (Deleuze, 1995, p. 175). This is what Deleuze calls societies of control, as I have discussed earlier. "Our ability to resist control, or submission to it, has to be assessed at the level of our every move. We need both creativity *and* a people" (Deleuze, 1995, p. 176, original emphasis). Nomadic education specifically acts upon this ideal.

### 1.2 *What Constitutes Education about Technology from a Deleuzian Perspective?*

In the preceding chapters I have attempted to demonstrate several issues that impact upon the ongoing process of becoming technologically literate. Important amongst these is that human beings have always had, and continue to have, a vital relationship with technology. This relationship has served to define the evolution of humans in concert with the evolution of technology. In other words, the process of becoming human occurs synchronously alongside the process of becoming technological and, crucially, vice versa. However, this process of becoming never reaches a final state. It is, rather, a metastable state, one that forms a continuous evolutionary process; one that might be described as being circular or helictical in nature. Further, becoming is always a process of becoming *in relation to other things: both organic and non-organic.* The becoming swimmer discussed in an earlier chapter, forms a relationship with the water in either the river, the sea or in a swimming pool, as well as with other possible things such as armbands, floats etc. No human can live and evolve devoid of a relationship with something other. This relational aspect of becoming is what has been described above as an assemblage, and for Deleuze, all assemblages reject the objective 'what *is*' or 'I *am*' or the verb to *be*, preferring instead to see the world through the lens of assemblages which are in "constant variation ... constantly subject to transformations" (Deleuze & Guattari, 2004, p. 90). These transformations are, thus, always in a process of differentiating between the stable 'what *is*' and the metastable 'what is yet to (be)come.'

Deleuze, in his vast corpus of work, rarely mentions education. He does, however, discuss at some length the concepts of thinking and learning, concepts that should underpin all forms of education but, which, according to Deleuze, do not. His reasoning for this is outlined in a chapter entitled 'The

Image of Thought' in his masterwork *Difference and Repetition.* He begins by differentiating between the concepts of recognition and imagination. Recognition is to know something that already exists, imagination is to formulate an idea, to think something new. Recognition relies upon "identifying something anew by comparing it with what is already known or what has already been experienced" (Williams, 2013, p. 127). In Chapter 5, I cited an article in the *Guardian* newspaper about artificial intelligence being able to learn independently. A French startup developed an artificial system called NooK which beat eight world champion players at the card game of Bridge. What was remarkable was that the AI won, not just by brute force of calculation, but by formulating rules. Moreover, it was able explain how its decisions were made. Whilst clever from an engineering basis, this does not, however, constitute learning according to Deleuze. Instead, it is a machine process that, as Williams suggested, above, is capable of "identifying something anew by comparing it with what is already known or what has already been experienced" (Williams, 2013). It is recognising patterns and developing a means to imitate, just as I was always told to do in my maths tests: always show your 'working out' because you will possibly get some marks for showing how you reached your answer, even if it is wrong. Nook would be more impressive if it were to show some emotion when it won, or to demonstrate its 'learning' by inventing a new version of bridge. AI still represents Friere's notion of data in/data out. There is no gain or indeed enjoyment in playing bridge with an emotionless machine. The concern in this for craft-based technology education is that the pedagogies relating to the development of AI are not too far removed from the pedagogies employed in craft-based technology education today: the pedagogies of knowing how and knowing that but not knowing why. Bogue (2008) summarises it thus:

> Thought's goal in a world of recognition and representation is to eliminate problems and find solutions, to pass from non-knowledge to knowledge. Learning in such a world is simply the passage from non-knowledge to knowledge, a process with a definite beginning and ending, in which thought, like a dutiful pupil, responds to pre-formulated questions and eventually arrives at pre-existing answers. (p. 7)

Thinking, according to Deleuze on the other hand, must move beyond the already given, beyond recognition and representation to the contemplation of the new. This can be brought about only by what he calls 'imageless thought.' Thinking, as part of the ongoing process of becoming ethnotechnologically literate, is thus disruptive to ordinary habits, "[f]or the new ... calls forth forces in thought which are not the forces of recognition, ... but the powers of a completely other model, from an unrecognised and unrecognisable *terra incognita*"[2] (Deleuze, 2004a, p. 172). Rather than eliminating problems by way of some pre-established solutions, Deleuze asks that we consider problems as being

> independent of the questions and solutions stored in memory. It is a problem because it does not yet have a solution and because it will never allow for solutions that cancels it out ... a problem is something inevitable ... that can only lead to a series of creative reactions rather than to a lasting solution. (Williams, 2013, p. 130)

This also applies to values. A nomadic pedagogy allows students to begin to recognise the difference between established values and the creation of new values relating to technology. Values, like everything else, are created by humans and so are constantly subject to transformation. New values about technology evolve and change, become established and are again subject to new perspectives. Established values decompose over time and recompose anew, much as Simondon's concept of pre-individuation and Deleuze's concept of the virtual.

The process of learning something new, for Deleuze, is not progressing from one state of not knowing what a specific tool is and not knowing how to use it, to another state of knowing what it is and how to use it. This would suggest that learning comprises a series of static states of knowing things that collectively appear to demonstrate a progression of knowledge, much like the static images in a movie film strip suggest movement. "According to Deleuze, that something new cannot simply be facts or skills since they are the result of moving from one state to another, rather than the explanation of how we go from state to new state" (Williams, 2013, p. 147). This model also implies that the required knowledges that students are expected to acquire in craft-based technology education, are prescribed in advance. This not only restricts any student agency in the process of 'learning,' as Deleuze implies, it simply makes no sense:

> We never know in advance how someone will learn: by means of what loves somone becomes good at Latin, what encounters make them a philosopher, or in what dictionaries they learn to think. The limits of the faculties are encased, one in the other, in the broken shape of that which bears and transmits difference. There is no more a method for learning than there is a method for finding treasures, but a violent training, a culture or *paideia*[3] which affects the entire individual. (Deleuze, 2004a, p. 205)

Learning is, thus, a personal project that is different for each individual. True learning is about experimenting with ideas, taking risks, becoming other. Deleuze argues that immanent becoming constitutes the affirmation of life where "[m]odes of life inspire ways of thinking; modes of thinking create ways of living. Life activates thought, and thought in turn affirms life" (2012, p. 66). Thinking, for Deleuze, is both a creative process and is political in nature. "When thought is genuinely *thinking* (and most of the time it is not), its nature is disruptive and hostile to the established strata of power" (Cisney, 2014, p. 37, original emphasis). Thinking, then, is a disruptive act. It is not simply reproducing stuff that is already known. That is just remembering. Thinking is the difference that occurs when individuals move from one mode of existence to another. This is when they become other. The political aspect of thinking about technology is made manifest in the process of *becoming*, or in Deleuze and Guattari's terms *deterritorialisation*, or in Guattari's terms *transversality*. Each of these concepts constitutes an act of disruption, a phase change that can be good or bad depending upon personal worldviews. "Nomadicism allows the maximum extension of principles and powers; if something can be thought, then no law outside thinking, no containment of thought within the mind of man should limit thinking's power" (Deleuze, 1994, p. 37). A nomadic philosophy of education about ethnotechnological literacy is a pedagogy of the power to disrupt and so create something other: a state of affairs that forms a new relationship between human beings and their technologies. This disruption will impact upon human beings in some way, whether good or bad. It must be remembered that technology is not neutral.

Nomadic education about technology starts from the acceptance that knowledge is not fixed but is created as part of an ongoing process. A nomadic education enables young people to be involved in the process of creating and critiquing alternative perspectives on what constitutes knowledge and meaning. It allows them to challenge each other's perspectives, even those of the teacher. This is what Nietzsche calls 'perspectivism.' For Nietzsche, "perspectival empathy encourages us to become tireless inquirers into the complexity, beauty and promise of other people's way of life … [his] doctrine of perspectivism thus yields a robust form of empathy with far-reaching consequences for civic interaction and engagement" (Jonas & Yacek, 2019, p. 179).

In nomadic education, the teacher's job is not to impart knowledge: it is to guide the process of co-creating knowledge with a view to becoming other. It offers an educational paradigmatic framework that helps guide young people towards the ability to facilitate their own personal and collective becoming other. By so doing, they can create their own new, exciting and sustainable possible futures. These futures may not be realised in actuality at present, but they will exist in a pre-individuated or virtual phase that has the potential to become realised, in some form, in the future. The very process of nomadic education helps raise awareness. Greta Thunberg and like-minded young activists serve as good examples of nomadic thinking: thinking that that does not let the ball and chain of politics, capitalism and consumerism get in the way of new possible futures.

The ongoing process of becoming ethnotechnologically literate, then, can be best achieved through a nomadic form of pedagogy. Becoming ethnotechnologically literate is more than ever before, a necessary cultural foundation enabling young people to critically analyse technology and develop an understanding of how technology has developed over specific periods and cultures. It is impossible to understand technology without understanding culture. It is, moreover, important to challenge the way the 'State' and mega-corporations utilise technologies of control, as I have discussed above.

## 2 Becoming Nomad in Education about Technology

> Article 1 of the Declaration of Human Rights states that human beings should act towards one another 'in a spirit of brotherhood.' This is the recognition that all others, all otherness, constitutes a community. (Tubbs, 2014, p. 121)

Becoming nomad, in terms of education about technology, is a progressive pedagogy that is distinct from the traditional forms of pedagogy normally associated with technology education. It is a form of education brought about through *experiencing* the world and is relevant for all, whether in school or beyond. "Even for a mind defined as immature, an open-ended world makes sense; it is real, and the child is empowered with the gift of assigning meaning to her being in this world, not by receiving the knowledge of it from some global viewpoint, but understanding and deriving it as contingent on her own local experience" (Semetski, 2006, p. 103). Regressive traditional forms of pedagogy employed within craft-based technology education classrooms today, assume knowledge, knowledge production and skills to be pre-existing and therefore stable, and subject to little or no change. Knowledges in this mode of technological classroom comprise repeatable skills and abilities that can be passed on from expert to novice. This constitutes a teacher-centred pedagogy which, as I have argued before, is at present the dominant and safest way to teach workshop orientated forms of knowledge and skill development. Paradoxically, however, although learning in this form of craft-based technology education claims to be creative and experiential, it is not. Even when students are allowed to design an artifact, the creative element is limited to the requirements of the imposed curriculum, the agreement of the teacher and whatever resources are available at the time. It is essentially hands-on-learning in which students participate, on an individual basis, under the guidance of the teacher. Even the design-folio format, when required, is largely prescribed. This, as I have discussed earlier, is also problematic because it does not take account of designing in the 'real' world. Students experience activities that have been pre-arranged for them by their teacher in association with the curriculum documentation that is current at the time. All learning is then subject to the tyranny of summative assessment in what Freire describes as the 'banking method' (1996). All meaning is, thus, considered to be fixed and permanent, rendering meaning absolute. There can only be right or wrong answers in this model. However, as I have argued above, knowledge is not fixed and immutable; it is created. Semetsky postulates further that, "[i]f no fixed and permanent meaning exists, then all knowledge is contingent, and such categories as reality, truth, knowledge, and the self as the subject of knowledge, become problematic" (Semetski, 2006, p. 92). In other words, given that meaning is created, it must be dependent for its existence upon the circumstances with which it finds itself associated. Technology, for example, as Ihde alludes, "allows us to see, to order, to relate to the world in a particular way" (1983, p. 33), a way that would have been inconceivable a hundred years ago, or even a couple of decades ago. Unlike the traditional model of technology education, teaching about technology or the ongoing process of becoming ethnotechnologically literate, requires a different pedagogical framework, one that a nomadic pedagogy can satisfy.

It is, however, important to understand that humans are complex subjects who see the world in different ways. To consider a technology education class as a single entity, a whole monolithic class that learns the prescribed content of a lesson in unison, is to deny that each individual student is complex and different.

Nomadic pedagogy does away with the tradition of an expert teacher passing on pre-existing knowledge collectively to a classroom of passive students. A quick reminder will help to highlight what technological literacy was thought to be in 2003:

> Citizens of today must have a basic understanding of how technology affects their world and how they can exist both within and around technology. The need for technological literacy is as fundamentally important to students as traditional core subject area knowledge and abilities.

> Students need and deserve the opportunity to attain technological literacy through the educational process. (Dugger & Wright, 2003, p. 2)

The nomadic pedagogy that I am proposing completely supports this statement of need. However, the ITEEA *Standards for Technological Literacy Revision Project* published in 2019 sought to revise, or at least upgrade the above notion of technological literacy because, they effectively argued, it was not able to be defined and assessed in a universal set of curriculum standards; "what constitutes [technological] literacy is a shifting landscape, subject to change as cultural conditions change ... The utility of such a literacy would depend on its ability to adapt and keep pace with constant change" (Buelin et al., 2019, p. 5). Citing Gagel they go on to argue

> Given that utility and adaptability depend heavily on what is known by the individual, the form of knowledge that appears most useful is praxiological[4] knowledge.' In response, Gagel laid out what he called an 'identity kit' for technological literacy – 'one containing those effectual elements having an inherent, unchanging, and enduring quality ... that would enable one to (a) accommodate and cope with rapid and continuous technological change, (b) generate creative and innovative solutions for technological problems, (c) act through technological knowledge both effectively and efficiently, and (d) assess technology and its involvement with human lifeworld judiciously. (Gagel, in Buelin et al., 2019, pp. 6–7)

Intriguing as this statement is, no mention is given, anywhere in the entire document, to pedagogy. Likewise in Dugger and Wright's citation earlier, there is only the brief statement: 'All teachers must be well prepared in pedagogy.' Herein lies one of the major problems in the delivery of technological literacy. Many definitions have been offered over the years as to what constitutes technological literacy, including in two books edited by myself. The problem lies, therefore, not merely in defining it, but in enabling it. Young people live in the world and are affected by it, just as we adults are. Schools, however, offer them no voice in this respect. Consequently, the concept of teaching technological literacy, although deemed to be important, lies dormant in the background of craft-based technology education. Issues relating to global warming, for example, are considered to be the purview the adult world. Young people do not believe they have any say in this, at least not in school. Greta Thunberg was given no official platform in the recent Cop26 conference in Glasgow. She was certainly given voice by the press and spoke to thousands of followers gathered in a public space in Glasgow, but the fact that she was not allowed to speak inside the main conference speaks volumes about the attitudes towards young people having a voice. When schools deny young people a voice, it should come as no surprise that they become disaffected. Nomadic pedagogy, on the other hand, gives voice to young people. It enables them to express themselves democratically, whilst understanding that opposing views are valid and must also be given voice. It allows for all perspectives to be heard and debated, but in a reasoned, conciliatory and constructive manner.

Borrowing from Kay Sidebottom's BERA presentation in 2019, she defines nomadic education thus:

> Nomadic [education] suggests a letting go of our old ways of thinking and becoming open to new ideas and modes of being. It is not resistance, but an understanding that we are always in the 'process of becoming' It requires an ethical standpoint that acknowledges the position of the [teacher and the students]: both *centering* themselves in an awareness of their biases, influences and their power in making 'agential cuts; … that present the findings which can only ever be limited by time, space and individual selection; and *decentering* themselves by acknowledging the influence and role of other agents in the learning assemblage (human and non-human others, material 'things,' the environment and so on). (Sidebottom, 2019)

'Decentering,' in this sense, is a willingness to step back from any deeply held view and become open to listening to the perspectives of others, some of which will represent opposing views. It does not mean that any individual necessarily gives up their particular views on some technologically orientated subject domain, but by listening to others they may, perhaps, wish, to some extent, to reformulate their perspective. Nomadic pedagogy about technology is, thus, less interested in procedural and declarative pedagogical structures. Rather "we should explore the conditions of the phenomena around us and focus not on the already made but on that which is in the making" (Sidebottom, 2019).

## 3 A Nomadic Pedagogy for Education about Technology

The skills and knowledges that are generated in craft-based technology education already exist, but are, for the most part, outside or external to the students lived experience. Learning is first organised on behalf of the students and then transmitted to them via the teacher. Essentially, the planned curriculum formulates specific State-directed, striated learning outcomes that are then imposed upon young people without their consent: because the curriculum says so! Education *about* technology, in contrast, is autopoietic[5] in nature. Knowledge and skills evolve as a direct result of the participants sharing their lived experiences. It is a self-organising curriculum of lived experience (Aoki, 2005). Learning is self-organised and immanent in nature.

> Because specialisation is defined as self-organising, it is accompanied by specific features that, in turn, affect the concept of learning which therefore becomes an emergent property of the transactions between teachers, students and the subject-matter, even in the absence of direct instruction and teaching as traditionally defined. (Semetski, 2006, p. 99)

What Semetsky means by specialisation is that the students have a say in what they want to debate and learn. They self-organise their own learning.

Self-organised learning is what Freire (1996) refers to as "liberating education" (p. 60). It occurs within the smooth space of multiple relationships where a variety of perspectives intersect and challenge each other, thereby making learning about technology more complex than in pedagogies of transmission. It creates more opportunities for intellectual growth. Nomadic learning creates an environment of perplexity, a novelty that requires choice. It requires participants to make their own decisions based upon their own experiences, variously blended with other perspectives, especially conflicting ones. These various perspectives, when considered together, have the potential to broaden out thinking about new technological possibilities by inventing solutions to scenarios that may not yet exist. The smooth space necessary for self-organised learning, means the formation of a space where the individual student's and teacher's beliefs and experiences meet. The varied worldviews about the technologically textured world collectively inhabited, will almost certainly conflict. However, where the teacher and the students agree that all are entitled to express what may be conflicting perspectives, and that perspectives may be challenged in a reasoned conciliatory and constructive manner, a nomadic pedagogy will have been created. The focus is no longer on providing the 'right' answer, with the concomitant fear of failure when a 'wrong' answer is given. Semetski (2006) argues that in a nomadic classroom environment, because of this type of change of focus, new and enriched experiences can evolve:

> among many available options [which] requires that those options are present but also stimulates the mode of thinking and acting so that students would not be horrified by possible contradictions and choices that may seem to oppose each other. Rather than perceiving a sense of failure,

> students … even in conflicting situations … may extract from them forces that vitalise the system by diversifying it, that is, by enriching [learning about technology] with *variations.* (p. 99, original emphasis)

This resonates to some extent with Martin Buber's concept of an I-Thou or an I-You relationship, where conflicting perspectives can discover common ground.

> According to Buber, an interpersonal encounter contains wonderful potential that far exceeds two separate people in conversation. This potential becomes apparent when two people actively and authentically engage each other in the here and now and truly "show up" to one another. In this encounter, a new relational dimension that Buber termed "the between" becomes manifest. When this between dimension exists, the relationship becomes greater than the individual contributions of those involved. This type of meeting is what Buber described as an I-Thou relationship.
>
> The I-Thou relationship is characterised by mutuality, directness, presentness, intensity and ineffability. Buber described the between as a bold leap into the experience of the other while simultaneously being transparent, present and accessible. He used the term "inclusion" to describe this heightened form of empathy. It is a far cry from the now-familiar scene of a group of friends sitting around a table at a restaurant, all gazing into their smartphones. (Martin & Cowan, 2019)

A nomadic pedagogy for learning *about* technology will inevitably involve conflicting perspectives which, under normal circumstances, remain resistant to other perspectives as shown in Figure 10.

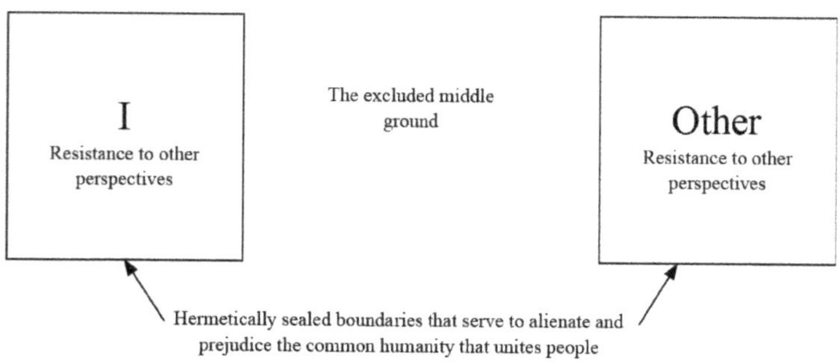


FIGURE 10 Buber's resistance to other perspectives

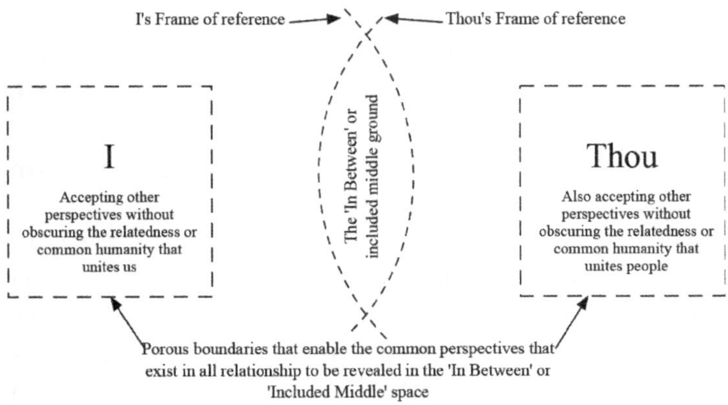


FIGURE 11 Buber's I-Thou relationship

However, by adopting Buber's I-Thou concept, which is not unlike Nietzsche's perspectivism or Deleuze and Gauttari's concept of territorialisation, deterritorialisation and reterritorialisation, or Guattari's transversality, new relationships, or even partial relationships can be made manifest, as indicated in Figure 11.

It is the variations that emerge in learning about technology, as a result of self-organised learning, that as Semetsky argues, enables a progressive pedagogy that deviates from the limited, one size fits all, whole class teaching paradigm that is most prevalent in craft-based technology education today. This model of nomadic education opens up access to an infinite range of perspectives, most of which, will challenge the status quo in sometimes quite surprising ways. As Goodchild (1996) suggests:

> For example, enemy soldiers confronting each other across a battlefield act out the contest that they are required to fight by distant political and economic entities, supported by nationalist, racist, or tribalist sentiments; their social relation is limited to an exchange of bullets, instead broadening out to become an exchange of jokes, songs, cigarettes, and memories. Revolution is not a matter of tearing up the script, forgetting or destroying external political and economic institutions, as well as internal conventions and expectations, for one is then left with no relation at all. Instead, revolution occurs through making additions to the script, bringing in unexpected amendments by borrowing strategies from elsewhere. The soldier who brandishes a pack of cards instead of a gun indicates possibilities for a different development of the plot. Deleuze and Guattari's writing is itself of a similar nature: instead of directly

> throwing aside theoretical norms, it offers a whole range of digressions and alternatives that carry thought elsewhere, shattering the coherence of hegemonic discourses. Liberation occurs through addition. (p. 2)

Surprising as it may seem, something not dissimilar to the scenario described above actually happened in Belgium on Christmas morning of 1914, at the beginning of the First World War. British and German troops, dug into opposing trenches, agreed to form a truce for the day. They played football and swapped gifts. Not unsurprisingly perhaps, the political and military hierarchies from both sides condemned this act as tantamount to treason. This provides an interesting, real-life example of a striated hierarchy attempting to occupy the smooth space created by the soldiers. Hence the need for Deleuze and Guattari's war machine.

Any act, or enquiry into the unknown, is nomadic if it "is experiential and experimental: the nomad's way is an immanent trajectory … But because for Deleuze social and psychic dimensions interpenetrate, nomadic ideas would be, from the epistemological perspective, intensive multiplicities distributed in smooth space" like the soldiers' collective experiment on Christmas day (Semetsky, 2009, p. 451). What is important to realise, is that for Deleuze, the process of becoming can take place only within smooth space. Whether it is the creation of art, philosophy or a technology, the process of individuation is conceptualised in the pre-individual or virtual phase and ends with the bringing into being of something that does not yet exist: a painting, a philosophical concept or a technology. However, as has been discussed earlier, Simondon's pre-individual phase and Deleuze and Guattari's virtual space, they precede actualisation. This is where the invention of concepts takes place. The emergence of a new concept must first be thought, created, invented, imagined or designed by human beings. Likewise, the affect that these concepts may currently have, or potentially could have on society, is open for debate in a nomadic classroom. The smoother this pre-individual/virtual space is, the more degrees of freedom to experiment are provided. The actualisation of technologies, however, is much more likely to require conformity to pre-existing rules and regulations in the real world of which they become a part. As the pre-individual/virtual technology in question transitions into becoming actual, the smooth space gives way to striated space. A technology may, for example, require electric power which will require a number of technological regulations to be incorporated into its formation. There may also be safety regulations, as well as many others, that must be adhered to. The design of buildings, products, aeroplanes, cars, in effect all technologies, are regulated to varying degrees when actualised. They must, consequently, conform to the striated, exterior regulations they will come to occupy. This is clearly reflected in the pedagogies employed in craft-based technology education today where regulated skill-based development takes precedence over expressive design. Philosophy, debate and art, on the other hand, are much less regulated and so have more degrees of freedom of expression in the pre-individual/virtual smooth space, as well as the potentially striated external spaces they may come to occupy. The transition from the pre-individual/virtual phase to actual concepts, is much less regulated for philosophy, art and debate and so can occupy more smooth space as part of the process of pre-individuation/virtuality. This is why a nomadic pedagogy of thinking about technology, which is the ongoing development of ethnotechnological literacy, requires the formation of smooth spaces to be made available in schools, in order to enable the bringing into being of concepts about technology that both exist and do not yet exist. For example, in a nomadic classroom, a discussion may be promoted regarding the creation of something that might be considered controversial, such as the digital world of the metaverse, as proposed by Mark Zuckaberg from Facebook. Unlike the conditions that may be imposed upon new technologies, this kind of research and debate can take place entirely in nomadic smooth space. These debates can then be widened to include other school cultures as will be discussed later in this chapter. Other debates, resulting from the ongoing process of becoming ethnotechnologically literate may include, for example:

- How can carbon emissions be reduced?
- What could we do personally to reduce global warming?
- What could be done to stop the deforestation of the Amazon rain forest?
- How can cyber bullying be stopped?
- Are there alternatives to electric cars?
- What would happen if there was no electric power for two weeks?
- What is technology?
- How does technology improve my life?

These debates are open-ended and have no pre-ordained solutions. They promote open ended democratic debate which enables student voice to be heard. This is in opposition to contemporary pedagogy in craft-based technological education that seeks the correct answers to questions asked. Dewey sums this up rather succinctly "There is all the difference in the world between having something to say and having to say something" (Dewey, 1907, p. 44).

The subjects of debates can be introduced by the teacher, or the students, or both. This format enables what Deleuze calls, "the fabulating a people in the process of becoming, or rather, a people-yet-to-come" (Carlin & Wallen, 2014,

p. XXI). Central to this pedagogy is the necessity to adopt a position of heterodoxy. Heterodoxy, as I define it, supports freedom of academic expression and idea generation where, in a technology education classroom setting, everyone, teachers and students together, are encouraged to discuss concepts relating to the technological past, present and possible futures. All ideas, past, present or even in terms of the future, are accepted no matter what. Deleuze refers to this as the creation of 'pas une idée juste, juste une idée' (not a correct idea, just an idea) The point he makes here, is as follows:

> just [or correct] ideas are always those that conform to accepted meanings or established precepts, they're always ideas that confirm something, even if it's something in the future of the revolution. While '[correct] ideas' is a becoming-present, a stammering of ideas [...] can only be expressed in the form of questions that tend to confound any answers. (Deleuze, 1995, pp. 38–39)

This pedagogy does not lend itself to offering clear, identifiable solutions to what are usually school based striated problems. Rather, a stammering of ideas interrogates past, present and potential future technological paradigms of expression. They are open to the development and creation of new and novel forms of technological expression. Problem solving in technology education today, tends to seek correct solutions, solutions that are seen to conform to certain pre-established criterion. A stammering of ideas, however, speculates upon a multitude of potentialities that may, hitherto, have been obscured from perception. Moreover, students will be encouraged to research issues, preferably in self-selected groups (see Freinet below).

Rather than working independently towards designing a problem that seeks a solution, nomadic thinking reorientates the process of thinking *about* technology. It does so by encouraging new forms of discourse that emphasise cooperative learning along the lines I have discussed above.

Teachers offering a nomadic pedagogy, however, need both support and assistance in how to manage learning in this model. This is also true for student teachers. Teaching under the rubric of nomadic education can be challenging. But, as Semetsky (2006) argues:

> What if teachers unreasonably, even if unintentionally, exceedingly maximise or neglectfully minimise the availability of alternatives by imposing some form of centralised control onto the classroom environment? Complexity, however, does not mean chaos. Contrary to centralised

> control and rule-based models, a self-organising system is … plastic and flexible. This means the dynamics proceed so that a system is capable of continuously adjusting – organising – itself in order to select that which is to be inhibited and that which is to be enhanced. Robustness and flexibility are two sides of the same coin. (p. 103)

This assumption is not too far removed from what a good teacher should be able to do in normal circumstances. However, given the rigidity of the curriculum today, as well as the expectations of governments about exam passes as indicators of success, it simply cannot be assumed that teachers will be able to teach ethnotechnological literacy under the auspices of a nomadic pedagogy, without some considerable support and assistance. Moreover, all educational stakeholders, from governments down to teachers themselves, need to change their attitudes about the purpose of education, perhaps by engaging more openly with crazy academics like me, in a nomadic line of enquiry. Without this change in attitude, the ongoing process of developing ethnotechnological literacy can only ever be enacted in a limited way, within the regulated structure of striated space. But, as I have already said, this reduces the likelihood of ethnotechnological literacy ever being taught in any creative and meaningful way. It needs an environment where collective free thinking about variations on human/technology – technology/human relationships can (be)come invented scenarios that do not yet exist. I offer below a highly praised example of nomadic education that, although conceptualised early last century, is just as relevant today. There are many other philosophical variations on progressive education that, whilst being shown to succeed, have never been adopted into mainstream education, providing yet another example of political hierarchies imposing striated curriculum content for technology education, designed to meet the perceived needs of industry.

## 4 Célestine Freinet (An Unwitting Educational Nomadologist)

> In opposition to logos, nomadology defines a way of thinking that, rather than rooting itself down in defence of one place or perspective, attempts to remain mobile and open to alterity and difference. (Haines, 2016)

Felix Guattari was very much influenced by the pedagogy devised by the French educationalist Célestine Freinet. Freinet (1896–1966), an educationalist and philosopher was considered to be the equal of Montessori, Piaget, Dewey, Freire, Giroux, Illich, Vygotsky, and Rousseau. He is, however, to this day, virtually unknown to those in the English-speaking world (Acker, 2007). He developed a methodology that put learning into the hands of the learners whilst simultaneously removing hierarchies, such as textbooks and rigid pedagogical structures designed to direct the flow of learning.

Thought about from the perspective of technology education, Freinet (1967) argued that:

> Like all social entities, the school must perforce adapt itself to the changing needs of the environment. Such adaptation is a fact of life: it can be observed even in the area of philosophy, which allows a kind of perpetual humanisation of technology and of life. Progress has always been redefined by the best among these thinkers. And thanks to them – at least to a certain extent – material development has been able to evolve into intellectual, moral, and human development. (pp. 100–101)

Freinet argued that young people could be better motivated if they had some say in their own learning. To this end, his philosophy of education can be summarised as follows:

- Student work must be productive and useful.
- Cooperative learning is necessary in the productive process.
- Group enquiry-based learning is based on trial and error [what Deleuze and Guattari call "experimentation"].
- The natural method is based on an inductive, global approach [now by way of abductive reasoning. See next chapter].

  Centres of interest are grounded in children's learning interests and curiosity (Cole, 2015, p. 87).

As a teacher, Freinet realised that anything he wrote on the chalkboard, which was often based upon classroom discussions, disappeared after he erased it. "There would be no record of this event in his students' life" (Acker, 2007, p. 13). This led him to acquire a small printing press, one that the students could learn to use in order to record their own learning. This became one of the two major initiatives that were the basis of his progressive and democratic form of education. The second major initiative he developed was the concept of interscholastic exchanges; the exchange of newsletters between schools that were written and printed by the students themselves.

These initiatives gave the students the impetus to engage in lively, critical classroom discussions, as well as enabling them to develop the more technical skills associated with printing. As part of the process, students were involved in developing their own learning.

Commenting in 1996 upon Freinet's method, Jean Haccuria, an inspector of schools in Brussels, summarised the basic tenets of Freinet's pedagogy. In a letter to the Belgian publication *L'Education Popular*, Haccuria pointed out some of Freinet's domains as including:

> free expression, free text, the printing in the school, freehand drawing, engraving on linoleum, free theatre, and current events (in Acker, 2007, p. 7). Moreover, "Freinet powerfully distrusted anything like a 'patent method' or a 'teaching formula.' He knew only too well how centrally imposed textbooks and courses of study could undermine teachers' best efforts to tap their pupils' natural [experiential] interest in regional events or activities that central planners would never take seriously." (Introduction, in Freinet, 1999, p. 3)

It should, therefore, be no great leap to articulate these initiatives into a thinking *about* technology paradigm. The printing press can be replaced by computer technology that will enable the writing of free texts to be made manifest electronically. The interscholastic exchanges can easily be accommodated through the Internet, providing all online exchanges are undertaken in a regulated safe space for young people.

## 5 Teaching Nomadic Education about Technology à la Freinet

Freinet believed that the teacher must act as facilitator, not as dictator. Logos must give way to nomos. The learning environment that Freinet inspired was one that facilitates, indeed encourages, the freedom necessary for the individual expression of perspectives about the world and the relationship of the individual with technology. Of crucial importance to this relationship, is how the individuals feel they are affected by technology and how they feel that relationship enables them to affect others. In order to facilitate this, teachers must recognise that while they "have an indispensable role to play, they should not monopolise classroom time since students should have a strong voice in classroom life" (Freinet, in Acker, 2007, p. 10).

Free texts enable learners to record, in both their own words and in their own style (which may include, sketching and photographing, for example), their personal and collective perspectives regarding their own and others' relationships with technologies. It is important that context is chosen by the learner and is not imposed from without. Classrooms, according to Freinet, should be part of the actual world and not a synthetic representation of the world. Free texts, then, can be used to aid each individual, in a cooperative setting, to express, freely, their recorded thoughts: thoughts that are focused upon issues relating to the way that they feel affected by technology and how they may affect others through the medium of technology. The teacher's role is to facilitate, not to impose.

> Freinet's educational method allows children to discover the vastness and the exigencies of freedom as it reduces what is 'forbidden in a classroom.' It allows them to choose a method of working: either an individual work plan or a collective work plan. Students discover real freedom – one that is not a whim or fancy but is engaged in a self-expression. (Acker 2007, p. 12)

Freinet proposed that interscholastic exchanges, mediated through the use of printed and illustrated school journals, written and formatted by the learners themselves, should form part of this new, nomadic, pedagogical framework. This facilitated the expansion of a young person's view of the world. Rural children, for example, could exchange information about technology with those living in urban or industrial areas. Today, with the advent of mass communication, the Internet can expand the scope and speed of this. Journals can be formatted online and sent electronically to other participating schools. This method can facilitate cultural exchanges where alternative perspectives about technology, technology education, and the ethical and political dimensions relating to technology can be revealed and further discussed.

One delightful example that sums up Freinet's progressive methodology is given in a record posted by a teacher in Algeria:

> One day, in my Second Grade classroom of 40 students, two of my young girls bring various texts on the Festival of the Sheep, and we very enthusiastically decide to prepare a document regarding the last night of Ramadan (known in Arabic as Eid El Kebir) to send to our correspondent in France. Quickly, in teams of two, the students divided the assignment: they wrote down what they knew, researched and questioned their parents and other elders on what they were unaware of. They filled out informational cards on the following subjects:
>
> – Origins of the festival
> – Purchase of the sheep and its arrival at the house
> – Slaughtering the sheep at the festival
> – Washing the sheep's fleece

> – Use of the meat and preparing the couscous
> – Various other dishes prepared on this occasion
> – Drying and storing the sheep's meat
>
> On the assigned day, each team presented the results of their research in front of an engrossed audience. After the students did all the necessary corrections, they carefully recopied the work, illustrated on cardboard the product of their research, put on a simple binding, adding a colourful cover and voila … their work was ready to be sent. As the teacher, what was my cost? Negligible. I directed their work a little and spent some of my personal time investigating the same thing they were investigating to make sure of its accuracy. I also got to know my students better. (In Acker 2007, pp. 46–47)

It is my belief that this methodology, which articulates with the philosophy of Deleuze and Guattari's concept of nomadism, can universalise the local. In other words, the concept of ethnotechnological literacy can become part of the various local technology education curricula around the world by, perhaps ironically, utilising modern technology as a platform that will enable students to express their various perspectives on technology in a global format. A modern adaptation of Freinet's methodology, using electronic journals and interscholastic exchanges, potentially including Zoom discussions, to facilitate freedom of expression, could provide the motivation for young people to engage with the technologically textured word they inhabit by sharing the outcome of their discussions with others who can comment and share in return.

The production of electronic journals, in whatever format is agreed between participating schools, can become guides, written and produced by the learners themselves, which give an account of their experiences, learning and deliberations regarding issues relating to technology and technology education. These guides, (not prescriptions), can be adapted, (not adopted), by future generations of learners. The journals, which can be made available to all in the participating schools as well as other interested parties such as parents or industry (young people influencing industry rather that the other way round), can be produced regularly, perhaps every month or two. Freinet found that because the journals were made available to a much wider audience than might otherwise be available, and produced on a regular basis, they had the effect of focusing the learners' attention on the production of high-quality work and enabled learning to take place over a wide range of subjects. Moreover, this learning took place on a voluntary basis, rather than by way of imposition.

An interesting, and more recent, example of this is a book entitled *Tomorrow is Too Late: An International Youth Manifesto for Climate Change*. The author/Editor is Grace Maddrell and the book was published in September of 2021 by the Indigo Press. It comprises 41 contributions from all around the world, provided by young people who give insights into the impact of climate change on their countries and themselves. Grace started to put the book together when she was just 15 years old.

Another book published in October 2021 by one boat Publishers is entitled *A Bigger Picture: My Fight to Bring a New African Voice to the Climate Crisis* by Vanessa Nakate who is 23 years old. Her book has been translated into French and German. These two examples demonstrate the power and influence that nomadic education about technology can have.

Assessment of ethnotechnological literacy can also form part of the nomadological approach. It cannot, however, under any circumstances, be assessed by summative examination. Given that nomadic pedagogy relies upon personal subjectivities and group dynamics, all of which will be different, the only form of assessment that is fair and reasonable is formative assessment, one in which students, teachers, and significant others contribute to the process. Moreover, nomadic education cannot, under any circumstances, be considered to be competitive. It can be guided only by means of cooperation and as such, any assessment must be (in)formative and continuous. In this way, any problems arising can be dealt with during the learning process rather than during some final examination. No one should pass or fail: students can only participate. Teachers should keep logs of any discussions held and conclusions reached. They will also have the electronic journals to evaluate. However, assessment can apply only to the class collectively, as is appropriate to the collaborative nature of the ongoing process of becoming ethnotechnologically literate. Any real assessment will be the influence these learners choose to exert in later life. This type of assessment is, fittingly, a nomadic form of assessment that supports the characteristics of a nomadic pedagogy for teaching about technology.

## Notes

1 Antinomy, contradiction, real or apparent, between two principles or conclusions, both of which seem equally justified; it is nearly synonymous with the term paradox.

2 A place, a subject, or situation that you are not familiar with.

3 Paideia: "– the classical Greek system of education and training, which came to include gymnastics, grammar, rhetoric, poetry, music, mathematics, geography, natural history, astronomy and the physical sciences, history of society and ethics, and philosophy – the complete

pedagogical course of study necessary to produce a well-rounded, fully educated citizen" (Tarnas, 1991, pp. 29–30).

4 Praxiological knowledge involves the study of human action.

5 Autopoietic is formed from *auto* (Greek prefix for self) and *poiesis* (Greek suffix meaning production, creation or formation).

CHAPTER 8

# Concluding Remarks and a Final Argument for a Progressive Education

I suggest that the nomadic approach, which may be considered as a pedagogy of freedom of expression, encourages young people to become more confident in expressing their own perspectives on issues important to them. These issues are of particular importance within the context of the exponential speed of technological development. Through this nomadic approach, an environment can be created in which learners are encouraged to work cooperatively on any endeavour relating to ethnotechnological literacy. This can be aided by removing the imposition of summative examination structures that focus only on the individual. The world of dualities (right or wrong, good or bad), gives way, in this model, to experimentation. There are no correct ideas, just ideas, as Deleuze argues. It is the kind of democratic education, as proposed by John Dewey, where new possible technological futures can be imagined; where abductive reasoning[1] becomes the natural method and the deductive reasoning of certainty is abolished. This can facilitate a learning space that encourages freedom of expression, whilst simultaneously leaving such views open to challenge. Any deeply held perspectives can be reasonably challenged by others, albeit in a reasoned, conciliatory and constructive manner. Students can learn to share and record their perspectives with others, the wider school and indeed, the world. They can be exposed to other cultures, other languages, and other alternative perspectives, potentially leading to changes in the ethnotechnologically textured world that they collectively inhabit. This can be achieved by the introduction of a nomadic pedagogy which, as described above, can include the production of digital journals shared with other schools across the globe. Perhaps, as an ongoing process of becoming ethnotechnologically literate, future generations will become more informed and better able to engage in critiques on gender, cultural, socioeconomic, global, and environmental issues.

I have researched some examples of progressive schools whose philosophy and pedagogical structures are eminently capable of sustaining this model for teaching technology, because they have instituted pedagogies which fit the needs of nomadic education in general. According to Rachel Sylvester (2021a), the *Sunday Times* education correspondent, the Südalinna school in Tallinn[2] Estonia is considered to be one of the best non-selective schools in Europe. "It has one of the best education systems in Europe, according to PISA which is

 | DOI:10.1163/9789004537002_008

run by the OECD"[3] (ibid.). Estonia operates on the principle of having a decentralised and equal education system, one in which the teachers decide what they teach and the pupils also have a say. Moreover, "teachers are encouraged to visit other schools and collaborate with each other" ... High on their educational outcomes list is "critical thinking, problem solving, communication and collaboration. Learning is about learning from each other and with each other. [They] say that it is OK to fail ... you can learn from that" (ibid.). Whilst not completely following what I am advocating precisely, there is more in common with my version of learning about technology than most state schools deliver. I have also mentioned earlier in the book, contemporary schools that are likewise, run along democratic principles. There is focus on holistic learning that involves learning from and with each other for example, as well as collaboration with other schools. There is also a rejection of a one-size-fits-all approach that is compatible with Freinet's philosophy.

Another example of a school that has a nomadically orientated approach is the XP School in Doncaster, England, which is a state funded school. This school has compassion, or a pedagogy that incorporates *sympathetic data*, embedded into its entire pedagogical framework. It considers the development of character (Bildung) and creativity to be as important as qualifications. Most radically, as outlined in another article by Rachel Sylvester, the subheading for the XP school[4] is as follows: "A State secondary school in Doncaster has ditched the traditional curriculum and given pupils a say on how they learn. The results? Outstanding" (2021b). Students learn 'real life' issues, or as Aoki (2005) suggests, learn through a curriculum of lived experience. According to Sylvester, the school is 11 times oversubscribed, and in 2018 Ofsted[5] judged XP to be outstanding in all categories. The school is also non-selective and the vast majority of pupils come from disadvantaged backgrounds. There is a strong focus on local learning. Public speaking also plays an important role in the culture of XP. This also fits with Freinet's method in which learning is designed to be shared with other schools, thus providing a way of communicating and sharing beyond the local. I have entitled this 'universalising the local.'

I support these modern educational initiatives that deliver a curriculum that is student led, by way of a nomadic pedagogy. However, whilst both schools serve as exemplary models of educational achievement, they are still funded by the state, which means that the state retains a high level of control. The degrees of freedom given to the XP school by the government has been exceptional, but it is still under the scrutiny of the Government. This means that things could very possibly change after the next election, as happened in New Zealand. Even at present, the XP school must prepare students for state imposed systems of summative assessment, which to all intents and purposes, guide curricular content. Indeed, school exceptionalism is measured on this basis.

These democratic schools remove hierarchical power systems and replace them with systems of student empowerment. Yet more examples of these progressive schools are Summerhill in England, where students have a vote on how the school is run. The Futurum school in Sweden "that does not have any classrooms, there is nothing like a timetable, no school bells can be heard, the students do not have any school bags, and teachers are without desks. Futurum, in particular, is a radical new school that "merges creative architecture with modern learning and teaching methods" (Deepak, 2005, p. 15). The Agora school in Roermond, The Netherlands, is another example, where there are no classrooms, no timetables, no curriculum and no separate year groups. The philosophy of the school, as espoused by the closest thing it has to a head teacher is "We want to have a joyful environment, but also an environment that raises questions, because that's what everything is about" (Sylvester, 2022). This echoes Braidotti's (2013) patronage of Spinoza in which "an ethically empowering relation to others aims at increasing one's *potentia* or empowering force and creates joyful energy in the process. The conditions that encourage such a quest involve the processes of transformation aiming at affirming positivity" (p. 343, original emphasis). Sylvester demonstrates the validity of this when she reports that last year 89 per cent of pupils at the Agora passed their diploma at or above their expected level.

These progressive schools operate on inclusive democratic methods "of teaching [that are] rooted in the understanding that everyone in a classroom brings their values, cultures, genders and ethnicities with them, and these have an impact on the learning experience" (Paguyo & Iturbe-LaGrave, 2021).

In terms of education about technology, nomadic pedagogy liberates participants from the exclusivity of pre-determined procedural and declarative knowledge domains that, I argue, is the dominant pedagogical orthodoxy in craft-based technology education. With this liberation, the rational epistemological world of 'what is' is transformed into a process ontological perspective of 'what might be?' 'How might one live?' This opens up innumerable possibilities for students to respond, by experimenting with ideas, which are unlikely to be ideal solutions, but are more likely to foster imaginative and creative learning. In this model, learners look for answers – plural, not the answer – singular. As an example, consider the highly novel ideas made manifest in the then, little known architectural association of Renzo Piano, Richard Rogers and Gianfranco Franchini. Together, they designed (not built) the Centre Pompidou in Paris, an admittedly controversial and experimental design in which the services that are normally hidden within a building have been reconceived as an exterior feature. You may love it, or you may hate it, but it is a revolutionary and exceptionally creative architectural masterpiece that, not only won the international architectural competition to build it, but went on to win the highest award in the architectural world: the Pritzker Prize. Had the competition committee decided to reject this entry, on the basis that its design did not follow establishment protocols, they might well have subsequently constrained these creative and experimental architects from going on to create some of the most magnificent buildings in the world today. Examples of these include the Lloyds building (Rogers) and the Shard (Piano), both in London.

To insist on certain protocols that constrain creativity is exactly what happened in the conventional academies of fine art in 19th century France. The new wave of impressionists from Monet to Cézanne were initially suffocated by the establishment who sought to reject their novel experimentations. Had they succeeded, we may never have experienced some of the most wonderful avant-garde art that shaped the 20th century.

In relation to the concept of constraint, Deleuze, again influenced by Spinoza, "considers an individual to be constrained when they remain cut off from their power of acting, when they remain slaves to external forces that subjugate their passions" (Smith, 2011, p. 125). Conversely, he considers an individual to be "liberated, or strong, or free when their power of acting increases, to the point where it produces active affections and ideas" (Smith, 2011, p. 125).

The main focus of this book has been on the development of a nomadic pedagogy to address issues related to teaching and learning in the ongoing process of becoming ethnotechnologically literate.

I have in this book, attempted to advance a new nomadic pedagogy for the ongoing process of becoming ethnotechnologically literate; a pedagogy that gives voice to student perspectives and concerns about the technologically textured world they inhabit. My arguments in this book demonstrate that the normal transmission pedagogies associated with craft-based technology education are not only impossible for the teaching and learning of ethnotechnological literacy, they contribute to the fact that this concept, that is considered, according to the literature, to be a vital component in education about technology, has never truly been incorporated into any craft-based technology education paradigm. This is compounded by the fact that while technological literacy is considered a necessary element in teaching about technology, the one element that never seems to get much attention is that of pedagogy. In the latest *Springer Handbook of Technology Education*, edited by Marc J. de Vries, there are approximately 90 mentions of the word pedagogy. I did not include any mentions in references. The majority of mentions in the text relate, moreover, to concerns about teachers' lack of understanding of pedagogies related to technology education. Many others are prefixed by adjectives such as 'integrated,' 'authentic,' or 'design orientated.' Others yet again are expressed in general terms such as 'appropriate pedagogies' or simply 'pedagogy.' Very few offer accounts of how these pedagogies might relate to the actual teaching of technological literacy and certainly do not relate to the teaching of ethnotechnological literacy.

I echo Inna Semetsky and Andrew Stables, by inserting my concept of ethnotechnological literacy into their more general theory relating to education. Ethnotechnological literacy, then:

> cannot be reduced to *a priori* learning objectives, aims, or right versus wrong answers for the purposes of various forms of measurement, for these are always reductions; education [about technology] as a whole is much richer than formal assessment often implies. Nor is learning [about technology] a merely individual matter; nor, for that matter, is it uniformly collective. Rather [ethnotechnological literacy is about the relationship between human beings and technology and is] embedded in relational dynamics, therefore continuously developing via a variety of interpretive ways. [Nomadic pedagogy about technology] is about interactions and relations: between students and teachers, between people and their environments, between ourselves and others and between elements of that which is to be learned, understood through various relational perspectives .... [Nomadic pedagogy about technology] insists that the material and the mental cannot be so clearly demarcated and that education [about technology] is not merely a matter of developing the mind in a narrow sense; nor is nature simply a mechanism. Rather, we are embodied creatures who live in rich networks of experience, and our phenomenal worlds overlap, just as human experience overlaps, but does not coincide, with the experiences of other sentient beings. (Semetski & Stables, 2014, pp. 1–2)

With that in mind, it is in this respect, that I would draw attention to an article I recently read that was first published in the Financial Times dated 2020, and later in Suzanna Arundhati Roy's book entitled *AZADI*, that offers a contemporary perspective highlighting the affect that the Coronavirus pandemic has/is having on our daily lives. This is a perspective that is entirely appropriate for the development of ethnotechnological literacy. Arundhati Roy, an acclaimed author and activist, wrote an article entitled *The Pandemic is a Portal*. I have reproduced the closing paragraphs which I believe resonate strongly with the rationale behind this book. It demonstrates that change can be achieved if those involved are united and ready to fight for it.

> Whatever it is, coronavirus has made the mighty kneel and brought the world to a halt like nothing else could. Our minds are still racing back and forth, longing for a return to "normality," trying to stitch our future to our past and refusing to acknowledge the rupture. But the rupture exists. And in the midst of this terrible despair, it offers us a chance to rethink the doomsday machine we have built for ourselves. Nothing could be worse than a return to normality.
>
> Historically, pandemics have forced humans to break with the past and imagine their world anew. This one is no different. It is a portal, a gateway between one world and the next.
>
> We can choose to walk through it, dragging the carcasses of our prejudice and hatred, our avarice, our data banks and dead ideas, our dead rivers and smoky skies behind us. Or we can walk through lightly, with little luggage, ready to imagine another world. And ready to fight for it. (Arundhati Roy, 2020a, 2020b, pp. 203–214)

Let us transpose Arundhati Roy's perspective by imagining a portal between craft-based technology education as it is delivered today and a new education *about* technology; one that is democratic and replete with a multitude of undiscovered potentialities; one in which everyone, and especially students, is involved; one in which collective and meaningful conversations can be held about possible future, technologically textured worlds in which new modes of existence that are sustainable and less exploitative of the planet can be considered. Would that not be something worth fighting for?

## Notes

1. Abductive conclusions are qualified as having a remnant of uncertainty or doubt, which is expressed in retreat terms such as "best available" or "most likely." Deductive arguments, on the other hand, are evaluated in terms of their validity and soundness. If all the premises are true, then the conclusion is true.
2. Available in English at https://www.tallinn.ee/est/sydalinna/In-English
3. PISA stands for Programme for International Student Assessment and OECD stands for Organisation for Economic Co-operation and Development.
4. Available at https://xpschool.org
5. Ofsted stands for Office for Standards in Education and is a United Kingdom Government body within the Department of Education.

# References


Acker, V. (2007). *The French educator Célestine Freinet. An enquiry into how his ideas shaped education.* Lexington Books.

Aoki, T. T. (2005). *Curriculum in a new key: The collected works of Ted T. Aoki* (W. F. Pinar & R. L. Irwin, Eds.). Lawrence Erlbaum Associates.

Aristotle. (2001). *The basic works of Aristotle* (R. McKeon, Ed.). Modern Library.

Aristotle. (2002). *Nichomachean ethics* (R. Crisp, Trans.). Cambridge University Press.

Arundhati Roy, S. (2020a, April 3). The pandemic is a portal. *Financial Times.*

Arundhati Roy, S. (2020b). *AZADI: Freedom; facism; fiction.* Penguin.

Axell, C. (2018). Technology and children's literature. In M. J. de Vries (Ed.), *International handbook on technology education* (pp. 895–911). Springer.

Bacon, F. (2019). *The advancement of learning* [eBook edition]. Good Press.

Barab, S. A., & Duffy, T. (2012). From practice fields to communities of practice. In D. Jonassen & S. Land (Eds.), *Theoretical foundations of learning environments* (pp. 29–65). Routledge.

Bardan, A. (2015). *Epistemology and political philosophy in Gilbert Simondon: Individuation, technics, social systems.* Springer.

Barlex, D. (2014). Enabling both reflection and action: A challenge facing technology education. In J. R. Dakers (Ed.), *New frontiers in technological literacy: Breaking with the past* (pp. 123–139). Palgrave Macmillan.

Barlex, D. (2018). Design and technology in England: An ambitious vision thwarted by uninteded consequences. In M. J. de Vries (Ed.), *International handbook of technology education* (pp. 109–124). Springer.

Barlex, D., & Banks, F. (2021). *Teaching STEM in the secondary school: Helping teachers meet the challenge* (2nd ed.). Routledge.

Barlex, D., & Steeg, T. (2017, May). *Re-building design and technology in the secondary school curriculum* [Working paper]. https://dandtfordandt.files.wordpress.com/2017/05/re-building-dt-v21.pdf

Barthélémy, J. H. (2012). Fifty key terms in the work of Gilbert Simondon. In A. De Boever, A. Murray, J. Roffe, & A. Woodward (Eds.), *Gilbert Simondon: Being and technology* (A. De Boever, Trans.) (pp. 203–231). Edinburgh University Press.

Barthes, R. (1964). The structuralist activity. *Partisan Review, 34,* 82–88.

Baudelaire, C. (1846). *Oeuvres completes.* Fusees, No. 22 (G. Croly, Trans.) [E-book]. Le Jeune Enchanteur. https://www.scribd.com/read/367421923/Charles-Baudelaire-Oeuvres-completes

Bauman, Z. (2003). *The individualized society.* Polity Press.

Beckmann, J. (1785). *Anleitung zur Technologie* [*Introduction to technology*]. Thomas Edlen von Tratten.

Bijker, W. E., Hughes, T. P., & Pinch, T. (2012). *The social construction of technological systems: New directions in the sociology and history of technology.* The MIT Press.

Blair, E. (2021, August 12). The Lorax warned us 50 years ago, but we didn't listen. *NPR.* Retrieved January 12, 2022, from https://www.npr.org/2021/08/12/1026385429/the-lorax-dr-seuss

Bogue, R. (2008). Search swim and see: Deleuze's apprenticeship in signs and pedagogy of images. In I. Semetsky (Ed.), *Nomadic education: Variations on a theme by Deleuze and Guattari* (pp. 1–16). Sense Publishers.

Bonta, M. (2013). 'We're tired of trees': Machinic university geography teaching after Deleuze. In I. Semetsky & D. Mansy (Eds.), *Deleuze and education* (pp. 57–73). Edinburgh University Press.

Braidotti, R. (2011). *Nomadic theory: The portable Rosi Braidotti.* Columbia University Press.

Braidotti, R. (2013). Nomadic ethics. *Deleuze Studies, 7*(3), 342–359.

Braidotti, R. (2019). *Posthuman knowledge* [eBook edition]. Polity Press.

Buelin, J., Daugherty, M. K., Hoepfl, M., Holter, C., Kelley, T., Loveland, T., … Sumner, A. (2019, June). *ITEEA standards for technological literacy revision project: Background, rationales and Structure.* ITEEA.org. https://www.iteea.org/File.aspx?id=151454&v=e868d0d8

Burke, J., & Ornstein, R. (1997). *The axmaker's gift: Technologies capture and control of our minds and culture.* Jeremy P Tarcher/Putnam.

Burnham, D., & Jesinghausen, M. (2010). *Nietzsche's Thus spoke Zarathustra.* Edinburgh University Press.

Butler, R. (2016). *Deleuze and Guattari's What is philosophy?* Bloomsbury.

Canguilhem, G. (1992). Machine and organ. In J. Crary & S. Kwinter (Eds.), *Incorporations* (M. Cohen & R. Cherry, Trans.) (pp. 44–69). Zone.

Carlin, M., & Wallen, J. (2014). Preface. In M. Carlin & J. Wallen (Eds.), *Deleuze and Guattari politics and education* (pp. XXI–XXVI). Bloomsbury.

Chabot, P. (2013). *The philosophy of Simondon: Between technology and individuation* (A. Krefetz & G. Kirkpatric, Trans.). Bloomsbury.

Chesterton, G. K. (2007). *Tremendous trifles.* Dover Publications.

Cisney, V. W. (2014, April). Becoming other: Foucault, Deleuze, and the political nature of thought. *Foucault Studies, 17*, 36–59.

Cole, D. R. (2017). Deleuze and learning. In M. A. Peters (Ed.), *Encyclopedia of educational philosophy and theory* (pp. 1898–1916). Springer Science+Business Media Singapore.

Cole, M. (1996). *Cultural psychology: A once and future discipline.* Harvard University Press.

Combes, M. (2013). *Gilbert Simondon and the philosophy of the transindividual* (T. LaMarre, Trans.). The MIT Press.

Comenius, J. A. (2010). *The great didactic of John Amos Comenius* (M. W. Keating, Trans.). Adam and Charles Black (FQ Legacy Books). (Original work published 1907)

Conference. (1868). *Conference on technical education held at Edinburgh.* Neil and Company.

Conley, T. (2010). Singularity. In A. Parr (Ed.), *The Deleuzian dictionary* (pp. 254–256). Edinburgh Univerist Press.

Dakers, J. R. (2005). Technology education as solo activity or socially constructed learning. *International Journal of Technology and Design Education*, 73–78.

Dakers, J. R. (2007a). Vocationalism: Friend or foe to design and technology education. In D. Barlex (Ed.), *Design and technology for the next generation* (pp. 90–107). Cliffeco Communications.

Dakers, J. R. (2007b). Incorporating technological literacy into classroom practice. In M. J. de Vries, R. Custer, J. R. Dakers, & G. Martin (Eds.), *Analysing best practices in technology education* (pp. 125–138). Sense Publishers.

Dakers, J. R. (2011). Activity theory as a pedagogical framework for the delivery of technology education. In M. Barak & M. Hacker (Eds.), *Fostering human development through engineering and technology education* (pp. 19–34). Sense Publishers.

Dakers, J. R. (2014a). *Defining technological literacy. Towards an epistemological framework.* Palgrave Macmillan.

Dakers, J. R. (2014b). *New frontiers in technological literacy.* Palgrave Macmillan.

Dakers, J. R. (2016). Taking a bicycle ride into the virtual with Simondon, Deleuze and Guattari. In M. C. van der Sanden & M. J. de Vries (Eds.), *Science and technology education and communication* (pp. 9–26). Sense Publishers.

Dakers, J. R. (2018). Nomadology: A lens to explore the concept of technological literacy. In M. J. de Vries (Ed.), *Handbook of technology education* (pp. 17–32). Springer.

Dakers, J. R., & De Vries, M. J. (2009). Perceptions of technology and technology education. In A. Jones & M. J. De Vries (Eds.), *International handbook on technology education* (p. 201). Sense Publishers.

Dakers, J. R., Dow, W., & McNamee, L. (2009). De-constructing technology's masculinity: Discovering a missing pedagogy in technology education. *International Journal of Technology and Design Education, 19*, 381–391.

Dakers, J. R., Hallström, J., & de Vries, M. J. (2019). *Reflections on technology for educational practitioners.* Brill | Sense.

Dasgupta, P. (2021). *The economics of biodiversity: The Dasgupta review.* HM Treasury.

DATA. (2021, August 7). *Blueprint 1000.* The Design and Technology Association. Retrieved August 7, 2021, from https://www.blueprint1000.org.uk

DATA. (2022, January 10). *For education.* The Design and Technology Association. Retrieved January 10, 2022, from https://www.data.org.uk/for-education/

De Montaigne, M. (2003). Chapter XII – Apology for Raimond Sebond. In M. De Montaigne & W. C. Hazlitt (Eds.), *The complete works of Michel De Mantaigne* (C. Cotton, Trans.) [E-book]. Everyman's Library.

de Vries, G. (2016). *Bruno Latour.* Polity Press.

de Vries, M. J. (1986). What do boys and girls think of technology? Pupils' attitudes towards technology. In J. H. Ratt & M. de Vries (Eds.), *Pupils' attitudes towards technology* (pp. 16–18). Eindhoven University of Technology.

de Vries, M. J. (2005). *Teaching about technology: An introduction to the philosophy of technology for non-philosophers*. Springer.

de Vries, M. J. (2006). Two decades of technology education in retrospect. In M. J. de Vries & I. Mottier (Eds.), *International handbook of technology education: Reviewing the past twenty years* (pp. 3–11). Sense Publishers.

de Vries, M. J. (2008). Gilbert Simondon and the dual nature of technical artefacts. *Society for the Philosophy of Technology eJournal, 12*, 23–35.

de Vries, M. J. (2010). Introducing van Riessen's work in the philosophy of technology. *Philosophia Reformata, 75*(1), 2–9.

de Vries, M. J. (2012). Utopian thinking in contemporay technology versus responsible technology for an imperfect world. *Perspectives on Science and Christian Faith, 64*(1), 11–19.

de Vries, M. J. (2018a). Technology education: An international history. In M. J. de Vries (Ed.), *International handbook of technology education* (pp. 73–84). Springer.

de Vries, M. J. (2018b). Philosophy of technology: Themes and topics. In M. de Vries (Ed.), *Handbook of technology education* (pp. 7–16). Springer International Publishing.

De Vries, M. J., & Tamir, A. (1997). Shaping concepts of technology: What concepts and how to shape them. *International Journal of Technology and Design Education, 7*, 3–10.

Deely, J. (2001). *Four ages of understanding: The first postmodern survey of philosophy from ancient times to the turn of the twenty-first century*. University of Toronto Press.

Deepak. (2005). Innovative teaching methods for the 21st century students. *I-manger's Journal on School Education, 1*(1), 15–19.

Deleuze, G. (1988). *Spinoza: Practical philosophy* (R. Hurly, Trans.). City Light Books.

Deleuze, G. (1995). *Negotiations* (M. Joughin, Trans.). Columbia Univesity Press.

Deleuze, G. (2002). On Gilbert Simondon. In G. Deleuze & D. Lapoujade (Eds.), *Desert islands and other texts, 1953–1974* (M. Taormina, Trans.) (pp. 86–89). Semiotext(e).

Deleuze, G. (2004a). *Difference and repetition* (P. Patton, Trans.). Continuum.

Deleuze, G. (2004b). *The logic of sense* (C. V. Boundas, Ed.; M. Lester & C. Stivale, Trans.). Continuum.

Deleuze, G. (2006). *Nietzsche and philosophy* (H. Tomlinson, Trans.). Continuum.

Deleuze, G. (2007). Responses to a series of questions. In R. Mackay (Ed.), *Collapse 3* (pp. 39–43). Urbanomic Media Ltd.

Deleuze, G. (2012). *Pure immanence: Essays on a life* (A. Boyman, Trans.). Zone Books.

Deleuze, G., & Guattari, F. (1994). *What is philosophy* (H. Tomlinson & G. Burchell, Trans.). Columbia University Press.

Deleuze, G., & Guattari, F. (2004). *A thousand plateaus: Capitalism and schizophrenia* (B. Massumi, Trans.). Continuum.

Deleuze, G., & Parnet, C. (1987). *Dialogues* (H. Tomlinson & B. Habberjam, Trans.). Columbia University Press. (Original work published 1977)

Deleuze, G., & Strauss, J. (1991). The fold. *Yale French Studies, 80*, 227–247.

Dennis, B., Kaplan, S., Bellack, M., Patel, K., & Adam, K. (2021, November 6). At COP26, 100,000 march for climate justice. *The Washington Post.* https://www.washingtonpost.com/climate-environment/2021/11/06/cop26-climate-protests-live-updates/

Derrida, J. (1987). Geschlect II: Heidegger's hand. In J. Sallis (Ed.), *Deconstruction and philosophy* (J. P. Leavey Jr., Trans.) (pp. 161–196). The University of Chicago Press.

Descartes, R. (2003). *Treatise of man* (T. S. Hall, Trans.) [E-book]. Prometheus Books.

Descartes, R. (2006). *A discourse on the method of correctly conducting one's reason and seeking truth in the sciences* (I. MacLean, Trans.). Oxford University Press.

Dewey, J. (1907). *The school and society.* University of Chicago Press.

Dewey, J. (1916). Democracy and education. In J. Boydston (Ed.), *John Dewey: The middle works, 1899–1924* (Vol. 9). Southern Illinois University Press.

Dow, W. (2006). The need to change pedagogies in science and technology subjects: A European perspective. *International Journal of Technology and Design Education, 16*, 307–321.

Dreyfus, H. L. (2001). *On the internet. Thinking in action.* Routledge.

Dugger, W. E., & Wright, M. D. (2003). *Advancing excellence in technological literacy: Student assessment, professional development, and program stadards.* International Technology Education Association.

Duignan, B. (1998, June 20). Apprenticeship. *Britannica.* Retrieved June 28, 2021, from https://www.britannica.com/topic/apprenticeship

Dunne, J. (1993). *Back to the rough ground: Practical judgement and the lure of technique.* Indiana University of Notre Dame Press.

Ellul, J. (1964). *The technological society.* Vintage Books.

Feenberg, A. (1999). *Questioning technology.* Routledge.

Feenberg, A. (2003). *Transforming technology.* Oxford University Press.

Feenberg, A. (2006). What is philosophy of technology? In J. R. Dakers (Ed.), *Defining technological literacy: Towards an epistemological framework* (pp. 5–16). Palgrave Macmillan.

Feenberg, A. (2014). What is philosophy of technology? In J. R. Dakers (Ed.), *Defining technological literacy: Towards an epistemological framework* (2nd ed., pp. 11–21). Palgrave Macmillan.

Fisher, R. (2000). *Teaching thinking: Philosophical enquiry in the classroom.* Continuum.

Franssen, M., Lokhorst, G. J., & van de Poel, I. (2018, September 21). Philosophy of technology. In E. N. Zalta (Ed.), *The Stanford encyclopaedia of philosophy*. Retrieved May 14, 2019, from https://plato.stanford.edu/archives/fall2018/entries/technology/

Freinet, C. (1967). *Education through work: A model for child centred learning* (J. Sivell, Trans.). The Edwin Mellen Press.

Freinet, C. (1999). *Cooperative learning and social change: Selected writings of Célestine Freinet* (D. Clandfield & J. Sivell, Trans.). Our Schools/Our Selves OISE Publishing.

Freire, P. (1996). *Pedagogy of the oppressed* (M. B. Ramos, Trans.). Penguin.

Geisel, T. (1999). *The lorex by Dr. Seuss.* Random House. (Original work published 1971)

Georgescu-Roegen, N. (2011). Energy and economic myths. In M. Bonaiuti (Ed.), *From bio economics to degrowth: Georgescu-Roegen's new economics in eight essays* (pp. 58–92). Routledge Studies in Ecological Economics.

Gidley, J. M., & Hampson, G. P. (2005). The evolution of futures in school education. *Futures, 37*(4), 255–271.

Giroux, H. (2020). *On critical pedagogy* (2nd ed.). Bloomsbury.

Golob, S. (2020). Exemplars, institutions, and self knowledge in Schopenhauer as educator. *Journal of Nietzsche Studies, 51*(1), 46–66.

Goodchild, P. (1996). *Deleuze and Guattari: An introduction to the politics of desire.* Sage Publications Kindle Edition.

Goodman, P. (2011). *New reformation: Notes of a neolithic conservative.* PM Press.

Government, H. M. (2021, August 7). *T-levels.* HM Government. Retrieved August 7, 2021, from https://www.tlevels.gov.uk

Gramsci, A. (1971). *Selections from the prison notebooks* (Q. Hoare & G. N. Smith, Trans.). International Publishers.

Gregory, A. (2008). Introduction. In Plato (Ed.), *Timaeus and Critias* (pp. IX–LIX). Oxford University Press.

Grigenti, F. (2018). Superpower and imitation body and machine. *Orbis Idearium, 6*(1), 13–38.

Grosz, E. (2004). *The nick of time: Politics, evolution, and the untimely.* Duke University Press.

Grosz, E. (2012). Identity and individuation: Some feminist reflections. In A. M. De Boever (Ed.), *Gilbert Simondon: Being and technology* (pp. 37–56). Edinburgh University Press.

Guattari, F. (1995). *Chaosmosis: An ethico-aesthetic paradigm* (P. Bains & J. Pefanis, Trans.). Indiana University Press.

Guattari, P. F. (2015). *Psychoanalysis and transversality: Texts and interviews 1955–1971* (A. Hodges, Trans.). Semiotext(e).

Hager, P., & Hyland, T. (2003). Vocational education and training. In N. Blake, P. Smeyers, R. Smith, & P. Standish (Eds.), *Philosophy of education* (pp. 271–287). Blackwell.

Hall, D. E. (2004). *Subjectivity* [eBook edition]. Routledge; Taylor and Francis Group.

Haraway, D. J. (1991). *Simians, cyborgs, and women: The reinvention of nature.* Free Association Books.

Hasse, C., & Wallace, J. (2020, September 29). *Towards a reconsideration of technological literacy.* Technucation. Retrieved January 19, 2021, from https://technucation.dk/en/concepts/technological-literacy/

Hayles, K. (2007). Hyper and deep attention: The generational divide in cognitive modes. *Modern Language Association Journal*, 187–199.

Hayles, N. (1999). *How we became posthuman: Virtual bodies in cybernetics, literature, and informatics.* University of Chicago Press.

Heidegger, M. (1972). *What is called thinking?* Harper and Row Publishers.

History.com Editors. (2010, February 9). *First transatlantic telegraph cable completed.* History. Retrieved January 14, 2020, from https://www.history.com/this-day-in-history/first-transatlantic-telegraph-cable-completed

Holland, E. W. (2003). *Deleuze and Guattari's A thousand plateaus: A readers' guide.* Bloomsbury Publishing Plc.

Holland, E. W. (2006). *Baudelaire and schizoanalysis: The sociopoetics of modernism.* Cambridge University Press.

Holland, E. W. (2011). *Nomad citizenship: Free market communism and the slow-motion general strike.* University of Minnesota Press.

Homer. (1952). *The Iliad of Homer and the Odyssey* (S. Butler, Trans.). Encyclopaedia Britannica Inc.

House of Lords. (2021, June 24). *Youth unemployment committee.* Retrieved August 7, 2021, from https://committees.parliament.uk/oralevidence/2440/html/

Hume, D. (2011). *Hume: The essential philosophical works: A treatise of human nature.* Wordsworth Editions Limited.

Huxley, A. (1959). *Brave new world revisited.* Chatto and Windus; Vintage. https://www.montclair.edu/iapc/what-is-philosophy-for-children/what-is-a-typical-p4c-session-like/

I.A.P.C. (2022, March 22). *What is a typical P4C session like?* Institute for the Advancement of Philosophy for Children: Montclair State University. Retrieved March 22, 2022, from https://www.montclair.edu/iapc/what-is-philosophy-for-children/what-is-a-typical-p4c-session-like/

IBO. (2021, January 20). *Design technology.* International Baccalaureate Organisation. Retrieved January 20, 2021, from https://www.ibo.org/programmes/diploma-programme/curriculum/sciences/design-technology/

Ihde, D. (1983). *Existential technics.* State University of New York Press.

Ihde, D. (1990). *Technology and the lifeworld: From garden to earth.* Indiana University Press.

Ihde, D. (1993a). *Philosophy of technology: An introduction.* Paragon.

Ihde, D. (1993b). *Postphenomenology: Essays in the postmodern context.* Northwestern University Press.

Ihde, D. (1997). The structure of technology knowledge. *International Journal of Technology and Design Education, 7*, 73–79.

Ihde, D. (2014). The designer fallacy and technological imagination. In J. R. Dakers (Ed.), *Defining technological literacy: Towards an epistemological framework* (2nd rev. updated ed., pp. 119–129). Palgrave Macmillan.

Ingerman, Ä., & Collier-Reed, B. (2011). Technological literacy reconsidered: A model for enactment. *International Journal of Technology and Design Education, 21*, 137–148.

ITEA. (2003). *Standards for technological literacy* (3rd ed.). International Technology Education Association.

ITEA. (2006). *Technological literacy for all* (2nd ed.). International Technology Education Association.

ITEA. (2007). *Standards for technological literacy* (3rd ed.). International Technology Education Association.

Jagodzinski, J. (2017a). A critical introduction to what is art education. In J. Jagodzinsky (Ed.), *What is art education? After Deleuze and Guattari* (pp. 1–62) [E-book]. Palgrave Macmillan.

Jagodzinski, J. A. (2017b). *What is art education? After Deleuze and Guattari.* Palgrave Macmillan.

Jamwal, B. S. (2012). Teacher education: Issues and their remedies. *International Journal of Educational Planning & Administration, 2*(2), 85–90.

Janaway, C. (2006). Plato and the arts. In H. H. Benson (Ed.), *A companion to Plato* (pp. 388–400). Blackwell Publishing.

Jonas, M. E., & Yacek, D. W. (2019). Nietzsche's philosophy of education rethinking ethics, equality, and the good life in a democratic age. In M. E. Jonas & D. W. Yacek (Eds.), *Nietzsche's philosophy of education* (p. 111). Taylor and Francis.

Jun, N. (2011). Deleuze, values, and normativity. In N. S. Jun (Ed.), *Deleuze and ethics* (pp. 89–107). Edinburgh University Press.

Kahn, C. H. (2004). *The art and thought of Heractilus: An edition of the fragments with translation and commentary.* Cambridge University Press.

Kant, E. (1998). *A critique of pure reason.* Cambridge University Press.

Kant, E. (2016). *An answer to the question 'what is enlightenment'.* Delphi Classics.

Kapp, E. (2018). *Elements of a philosophy of technology: On the evolutionary history of culture* (J. W. Kirkwood & L. Weatherby, Eds.; K. K. Wolfe, Trans.). University of Minnesota Press. (Original work published 1877)

Keirl, S. (2015). 'Seeing' and 'interpreting' the human-technology phenomenon. In P. J. Williams, A. Jones, & C. Buntting (Eds.), *The future of technology education* (pp. 13–34). Springer.

Keirl, S. (2018). Design and technology education and its curriculum challenges. In M. J. de Vries (Ed.), *International handbook of technology education* (pp. 219–233). Springer.

Kirk, M. (2014). Reenvisioning our knowledge tradition: From gender blind to gender aware. In J. R. Dakers (Ed.), *New frontiers in technological literacy: Breaking with the past* (pp. 77–88). Palgrave Macmillan.

Koch, G. (2019). Animation of the technical and the quest for beauty. In T. Pringle, G. Koch, & B. Stiegler (Eds.), *Machine: In search of media* (pp. 1–23). University of Minnesota Press and Meson Press.

Kolb, D. A. (1984). *Experiential learning: Experience as the source of learning and development.* Prentice Hall.

Kramer, S. M. (1956). *History begins at Sumer.* University of Pennsylvania Press.

Krell, D. F. (2002). *Basic writings Martin Heidegger.* Routledge.

Kroes, P. (2002). Design methodology and the nature of technical artefacts. *Design Studies, 23*(3), 287–302.

Kroes, P., & Meijers, A. (2000). The dual nature of technological artefacts. *Studies in the History of Philosophy of Science, 37*, 1–4.

Kroes, P., & Meijers, A. (2006). Introduction. In P. Kroes & A. Meijers (Eds.), *The dual nature of technical artefacts* (Vol. 37, pp. 1–4). Elsevier.

Kuhn, T. (2012). *The structure of scientific revolutions* (4th ed.). The University of Chicago Press.

La Mettrie, J. O. (2016, May 16). *The project Gutenberg ebook of man a machine* (M. W. Calkins, Ed.; R. A. Watson, & M. Rybalka, Trans.). Gutenberg.org. Retrieved August 10, 2020, from https://www.gutenberg.org/files/52090/52090-h/52090-h.htm

Larcher, J. (2011). *Masters of cinema: Charlie Chaplin.* Cahiers du Cinéma.

Latour, B. (2005). *Reassembling the social.* Oxford University Press.

Lave, J. (1993). Situated learning in communities of practice. In L. B. Resnick, J. M. Levine, & S. D. Teasley (Eds.), *Perspectives on socially shared cognition* (pp. 17–36). American Psychological Association.

Lemke, J. (1997). Cognition, context and learning. A social semiotic perspective. In D. Krishner & J. A. Whitson (Eds.), *Situated cognition, social, semiotic, and psychological perspectives* (pp. 37–56). Lawrence Erlbaum Associates.

Leroi-Gourhan, A. (1993). *Gesture and speech* (A. B. Berger, Trans.). Massachusetts Institute of Technology.

Lipman, M., Sharp, A. M., & Oscanyan, F. S. (1980). *Philosophy in the classroom* (2nd ed.). Temple University Press.

Lord, B. (2010). *Spinoza's ethics.* Edinburgh University Press.

Lotka, A. J. (1925). *Elements of physical biology.* William and Wilkins Company.

Lyotard, J. F. (1991). *The inhuman* (G. Bennington & R. Bowlby, Trans.). Polity Press.

Martin, M., & Cowan, E. W. (2019, May 8). Remembering Martin Buber and the I-Thou counselling. *Counselling Today*. Retrieved March 30, 2022, from https://ct.counseling.org/2019/05/remembering-martin-buber-and-the-i-thou-in-counseling/

Marx, K. (1990). *Capital: A critique of political economy; Volume 1* (B. Fowkes, Trans.). Penguin Classics.

Massumi, B. (2012). 'Technical mentality' revisited: Brian Massumi on Gilbert Simondon. In A. De Boever, B. Murray, J. Roffe, & A. Woodward (Eds.), *Gilbert Simondon: Being and technology* (pp. 19–36). Edinburgh University Press.

Matusov, E. (1998). When solo activity is not privileged: Participation and internalisation models of development. *Human Development, 41*, 326–349.

May, T., & Semetsky, I. (2008). Deleuze, ethical education, and the unconscious. In I. Semetsky (Ed.), *Nomadic education: Variations on a theme by Deleuze and Guattari* (pp. 143–158). Sense Publishers.

Mayor, R. E. (1992). Cognition and instruction: Their historic meaning within educational psychology. *Journal of Educational Psychology, 7*(1–2), 141–159.

Michael, M. (2014). How to understand mundane technology: New ways of thinking about human-technology relations. In J. R. Dakers (Ed.), *Defining technological literacy: Towards an epistemological framework* (2nd ed., pp. 41–56). Palgrave Macmillan.

Mills, S. (2016). *Gilbert Simondon: Information, technology and media.* Rowman and Littlefield International.

Mitcham, C. (1994). *Thinking through technology: The path between engineering and philosophy.* The University of Chicago Press.

Mitcham, C. (2010). Placing technology in religious-philosophical perspective: A dialogue among traditions. *Philosophia Reformata, 75*(1), 10–35.

Moravec, H. (1988). *Mind children: The future of robot and human intelligence.* Harvard University Press.

Mumford, L. (2003). Tool-users vs. homo sapiens and the megamachine. In R. Scharff & V. Dusek (Eds.), *Philosophy of technology: The technological condition* (pp. 334–351). Blackwell Publishing.

Munroe, W. S. (1900). *Comenius and the beginnings of educational reform.* Charles Scribner's Sons.

Murphie, A., & Potts, J. (2003). *Culture and technology.* Palgrave Macmillan.

Murphy, P. (2006). Gender and technology: Gender mediation in school knowledge construction. In J. R. Dakers (Ed.), *Defining technological literacy: Towards an epistemological framework* (pp. 219–237). Palgrave Macmillan.

Murphy, P. (2007). Gender and technology. In D. Barlex (Ed.), *Design and technology for the next generation* (pp. 237–251). TEP and Cliffeco.

Naussbaum, M. (1979). Eleatic conventionalism and Philolaus on the conditions of thought. *Harvard Studies in Classical Philology, 83*, 63–108.

Nietzche, F. (1968). *The will to power* (W. Kaufmann, Ed.; W. Kaufmann & R. J. Hollingdale, Trans.). Vintage Books.

Nietzsche, F. (1911). *The complete works of Friedrich Nietzsche: Ecce homo: Nietzsche's autobiography* (Vol. 17) (O. Levy, Ed.; A. M. Ludovici, Trans.). The Macmillan Company.

Nietzsche, F. (2003). *Thus spoke Zarathustra* (R. J. Hollingdale, Trans.). Penguin Classics.

Nietzsche, F. W. (2014). *Schopenhauer as educator* (D. Pellerin, Ed.; D. Pellerin, Trans.). Creative Space Independent Publishing Platform.

Nietzche, F. (2010). *Thus spoke Zarathustra* (B. Chapko, Ed.) [eBook edition]. Feed Books.

OED. (1987). *The compact edition of the Oxford English dictionary supplement* (R. W. Burchfield, Ed.). Clarendon Press.

Oldenziel, R. (1999). *Making technology masculine: Men, women and modern machines in America.* Amsterdam University Press.

O'Riley, P. A. (2003). *Technology, culture, and socioeconomics: A rhizoanalysis of educational discourses.* Peter Lang.

Paetcher, C. (2007). *Being boys being girls: Learning masculinities and femininities.* Open University Press.

Paguyo, C. H., & Iturbe-LaGrave, V. (2021, April 3). *Making teaching as inclusive as possible.* Futurum: Inspiring the Next Generation. Retrieved April 3, 2021, from https://futurumcareers.com/making-teaching-as-inclusive-as-possible

Paterson, L. (2003). *Scottish education in the twentieth century.* Edinburgh University Press.

Pearson, G., & Young, A. (2006). *Why all Americans need to know more about technology.* National Academy Pres.

Petrina, S. (2007). *Advanced teaching methods for the technology classroom.* Information Science Publishing.

Petrina, S. (2018). From crit to social critique. In M. J. de Vries (Ed.), *International handbook on technology education* (pp. 39–50). Springer.

Petrina, S. (2020). Philosophy of technology for children and youth. In D. Barlex & P. J. Williams (Eds.), *An international perspective on pedagogy for technology education in secondary schools* (pp. 311–323). Springer.

Pierce, C. S. (1998). *The essential Pierce: Selected philosophical writings* (Vol. 2). Indiana University Press.

Plato. (2008). *Timaeus and Critias* (R. Waterfield, Trans.). Oxford University Press.

Plato. (2011a). *Meno* (J. Benjamen, Trans.) [eBook edition]. Neeland Media LLC.

Plato. (2011b). *Phaedrus* (B. Jowett, Trans.) [eBook edition]. Digireads.Com Publishing.

Plato. (2012). *Cratylus* (B. Jowett, Trans.). Start Publishing LLC.

Plato. (2013). *Symposium* (B. Jowett, Trans.). Project Gutenberg EBook.

Plotnisky, A. (2009). Bernhard Riemann. In G. R. Jones (Ed.), *Deleuze's philosophical lineage* (pp. 190–208). Edinburgh University Press.

Porkorny, J. (1967). *Indogermanishes etymologisches Wörterbuch.* Franke.

Prigogene, I., & Stengers, E. (1986). *Order out of chaos: Man's new dialogue with nature.* William Collins and Sons.

Protevi, J. (2001). *Political physics: Deleuze, Derrida and the body politic.* The Athlone Press.

Proust, M. (2003). *In search of lost time. Volume V1: Time regained* (A. Mayor & T. Kilmartin, Trans.). The Modern Library.

Pye, D. (1968). *The nature and art of workmanship.* The Herbert Press.

Pye, D. (1999). *The nature & aesthetics of design.* The Herbert Press.

Rani, A. (2017). Problems and solution of teacher education. *International Journal of Academic Research in Education and Review, 5*(1), 15–19.

Rasinen, A. (2003). An analysis of the technology education curriculum of six countries. *Journal of Design and Technology Education, 15*(1), 31–47.

Rockström, J., & Gaffney, O. (2021). *Breaking boundaries: The science of our planet.* Dorling Kindersley.

Roffe, J. (2010). Multiplicities. In A. Parr (Ed.), *The Deleuzian dictionary; Revised edition* (pp. 181–182). Edinburgh University Press.

Roochnik, D. (1996). *On art and wisdom: Plato's understanding of techne.* Pennsylvania State University Press.

Rousseau, J. J. (1979). *Emile or on education* (A. Bloom, Trans.). Basic Books.

Rousseau, J. J. (2002). *The social contract and the first and second discourses.* Yale University Press.

Russell, H. B., & Pertti, J. P. (1972). *Technology and social change.* Macmillan Co.

Ryle, G. (2000). *The concept of mind.* Penguin Books.

Sauvagnarges, A. (2016). *Artmachines: Deleuze, Guattari, Simondon* (S. Verderber & E. Holland, Trans.). Edinburgh University Press.

Sawday, J. (2007). *Engines of the imagination: Renaissance culture and the rise of the machine.* Routledge.

Schatzberg, E. (2018). *Technology: Critical history of a concept.* University of Chicago Press.

Schrödinger, E. (2013). *What is life: The physical aspects of the living cell.* Cambridge University Press.

Selfe, C. L., & Burns, H. (1999). *Technology and literacy in the 21st century: The importance of paying attention.* Project MUSE. Southern Illinois University Press.

Seller, R. M. (2021, June 28). *Semiology // semiotics.* UCalgery. Retrieved June 28, 2021, from http://people.ucalgary.ca/~rseiler/semiolog.htm

Semetski, I. (2006). *Deleuze, education and becoming.* Sense Publishers.

Semetsky, I. (2008). *Nomadic education: Variations on a theme by Deleuze and Gauttari.* Sense Publishers.

Semetsky, I. (2009). Deleuze as a philosopher of education: Affective knowledge/effective learning. *The European Legacy, 14*(4), 443–456.

Semetsky, I. (2015). Reading Kristeva through the lens of edusemiotics: Implications for education. *Educational Philosophy and Theory, 47*(10), 1069–1081.

Semetsky, I. (2017). Introduction: A primer on edusemiotics. In I. Semetsky (Ed.), *Edusemiotics: A handbook* (pp. 1–14) [eBook edition]. Springer.

Semetsky, I. (2020). *Semiotic subjectivity in education and counselling* [eBook edition]. Routledge.

Semetski, I., & Stables, A. (2014). *Pedagogy and edusemiotics: Theoretical challenges/practical opportunities.* Sense Publishers.

Shavero, S. (2014). *The universe of things: On speculative realism.* University of Minnesota Press.

Shaviro, S. (2006, January 16). Simondon on individuation [Blog]. The Pinocchio Theory. Retrieved June 19, 2020, from http://www.shaviro.com/Blog/?p=471

Sidebottom, K. (2019, September 2019). *Nomadic enquiry: Reimagining research ethics for posthuman lives.* BERA. Retrieved January 23, 2022, from https://www.bera.ac.uk/blog/nomadic-enquiry-reimagining-research-ethics-for-posthuman-times

Simondon, G. (1995). *L'individu et sa genése physico-biologic* (J. Dakers, Trans.). Éditions Jérôme Millon.

Simondon, G. (2003). *L'individuation à la lumiére des notions de form et d'information* (2nd ed.). Éditions Jérôme Millon.

Simondon, G. (2017). *On the mode of existence of technical objects* (C. Malaspina & J. Rogove, Trans.). Univocal.

Simondon, G. (2020a). *L'Individuation à la lumiére des notions de forme et d'information* [Individuation in light of notions of form and information] (Vol. 1) (T. Adkins, Trans.). Univesity of Minnesota Press.

Simondon, G. (2020b). *Individuation in light of notions of form and information* (Vol. 2) (T. Adkins, Trans.). Univesity of Minnesota Press.

Slattery, P. (2006). *Curriculum development in the postmodern era.* Routledge.

Smith, D. W. (2011). Deleuze and the question of desire: Towards an immanent theory of ethics. In N. S. Jun (Ed.), *Deleuze and ethics* (pp. 123–141). Edinburgh University Press.

Smith, D. W. (2012). Concepts and creation. In R. Braidotti & P. Pisters (Eds.), *Revising normativity with Deleuze* (pp. 175–188). Bloomsbury.

Smith, D. W. (2012). Deleuze and Derrida, immanence and transcendence: Two directions in recent French thought. In D. Smith (Ed.), *Essays on Deleuze* (pp. 271–286). Edinburgh University Press.

Smith, D. W., & Protevi, J. (2020, March 21). Gilles Deleuze. In E. N. Zalta (Ed.), *Stanford encyclopedia of philosophy.* Retrieved November 15, 2020, from https://plato.stanford.edu/archives/spr2020/entries/deleuze/

Snow, C. P. (1998). *Two cultures.* Cambridge University Press.

Souriau, É. (2015). *The different modes of existence* (E. Beranek & T. Howles, Trans.). Univocal Publishing.

Spengler, O. (2015). *Man and technics: A contribution to a philosophy of life* (M. Putnam & J. B. Morgan, Eds., C. F. Atkinson & M. Putman, Trans.). Arktos Ltd.

Spinoza, B. (1955). *On the improvement of the understanding/the ethics/correspondence* (R. Elwes, Trans.). Dover Publications.

Spinoza. (2000). *Spinoza: Ethics* (G. H. Parkinson, Ed.; G. H. Parkinson, Trans.). Oxford University Press.

Stagoll, C. (2010a). Becoming. Difference. In A. Parr (Ed.), *The Deleuze dictionary: Revised edition* (pp. 25–27, 74–76). Edinburgh University Press.

Stagoll, C. (2010b). Concepts. In A. Parr (Ed.), *The Deleuze dictionary* (pp. 53–55). Edinburgh University Press.

Stiegler, B. (1998). *Technics and time, 1: The fault of epimetheus* (R. Beardsworth & G. Collins, Trans.). Stanford University Press.

Stiegler, B. (2010a). *Taking care of youth and the generations* (S. Barker, Trans.). Stanford University Press.

Stiegler, B. (2010b). Memory. In W. J. Mitchell & M. Hansen (Eds.), *Critical terms for media studies* (pp. 64–87). University of Chicago Press.

Stiegler, B. (2016). *Automatic society: Volume 1; The future of work* (D. Ross, Trans.). Polity.

Stoltz, S. (2021). Nietzsche, eternal recurrence and education: The role of the great cultivating thought in the art of self-cultivation (Bildung). *Journal of the Philosophy of Education, 55*(1), 186–203.

St. Pierre, E. (2008). Deleuzian concepts for education: The subject undone. In I. Semetsky (Ed.), *Nomadic education: Variations on a theme by Deleuze and Guattari* (pp. 183–196). Sense Publishers.

Strum, S., & Latour, B. (1999). Redefining the social link: From baboons to humans. In D. MacKenzie & J. Wajcman (Eds.), *The social shaping of society* (pp. 116–124). Open University Press.

Svenningsson, J. (2019). Carl Mitcham: Descriptions of technology. In J. R. Dakers, J. Hallström, & M. J. de Vries (Eds.), *Reflections on technology for educational practitioners: Philosophers of technology inspiring technology education* (pp. 13–24). Brill | Sense.

Sykes, G., Bird, T., & Kennedy, M. (2010). Teacher education: Its problems and some prospects. *Journal of Teacher Education, 61*(5), 464–476.

Sylvester, R. (2021a, November 6). Want the best school in Europe? Try Estonia. *The Times/Sunday Times: Electronic Version.* https://www.thetimes.co.uk/article/024dab06-3bda-11ec-a9ce-48a11f44f00d?shareToken=9315e69844c2907a03af4cb482a34a4b

Sylvester, R. (2021b, October). Is XP the school of the future. *The Times/Sunday Times: Electronic Version.* https://www.thetimes.co.uk/article/no-uniforms-no-detentions-no-maths-lessons-is-xp-in-doncaster-the-school-of-the-future-jnc6n80kq

Sylvester, R. (2022, January 8). Would you send your children to a school with no rules? *The Times.* Retrieved January 8, 2022, from https://www.thetimes.co.uk/article/would-you-send-your-children-to-a-school-with-no-rules-x9pvchc5d

Tampio, N. (2010, August 30). Multiplicity. In *Encyclopedia of political theory.* https://faculty.fordham.edu/tampio/Tampio%20-%20Multiplicity.pdf

Tarnas, R. (1991). *The passion of the western mind: Understanding the ideas that have shaped our world view.* Ballantine Books.

Todd, R. D. (1991). The nature and challenges of technological literacy. In M. J. Dyrenfurth & M. R. Kozak (Eds.), *Technological literacy* (pp. 10–27). Glencoe, McGaw-Hill.

Toft, P. (2007). Design and technology: Seeing both the wood and the trees. In D. Barlex (Ed.), *Design and technology for the next generation* (pp. 266–294). Cliffeco Communications.

Tubbs, N. (2014). *Philosophy and modern liberal arts education: Freedom is to learn.* Palgrave Macmillan.

UNESCO. (2020, March 23). *Education for sustainable development: A roadmap: ESD for 2030.* Retrieved March 23, 2021, from https://unesdoc.unesco.org/ark:/48223/pf0000374802.locale=en

Unknown. (2008). *Exploitation of whales.* Marine Mammal Institute, Oregon State University. Retrieved January 12, 2022, from http://faculty.valenciacollege.edu/tklenk/marinecd/labs/PDFLABS/mammallabs/whale%20exploitation%20lab.pdf

Unknown. (2020, May 11). *Trois femmes (Le Grand Déjeuner), Fernand Léger: Analyse.* Gallerix Online Museum. Retrieved May 11, 2020, from https://fr.gallerix.ru/pedia/paintings-analysis--three-women/

van der Laan, J. M. (2016). *Narratives of technology.* Palgrave Macmillan.

van Manen, M. (2015). *Pedagogical tact: Knowing what to do when you don't know what to do.* Routledge.

Van Riessen, H. (1952). *The society of the future* (D. H. Freeman, Trans.). Presbyterian and Reformed Publishing Co.

Vanderstraeten, R. (2004). Education and society: A plea for a historicised approach. *Journal of Philosophy of Education, 38*(2), 195–206.

Vernant, J. (2006). *Myth and thought among the Greeks* (J. Lloyd & J. Fort, Trans.). Zone Books.

Vygotsky, L. S. (1978). *Mind in society. The development of higher psychological processes.* Harvard University Press.

Vygotsky, L. S. (1993). *The collected works of L. S. Vygotsky, Volume 2: The fundamentals of defectology (abnormal psychology and learning disabilities* (R. W. Reiber & A. S. Carton, Eds.; J. Knox, Trans.). Plenum Publishers.

Walkerdine, V. (1997). Redefining the subject in situated cognition theory. In D. Krishner, & J. A. Whitson (Eds.), *Situated cognition, social, semiotic, and psychological perspectives* (pp. 57–70). Lawrence Erlbaum Associates.

Wallen, J. (2013). Morphologies for a pedagogical life. In I. Semetsky & D. Mansy (Eds.), *Deleuze and education* (pp. 196–214). Edinburgh University Press.

Webster, R. S. (2009). Why educators should bring an end to pedagogy. *Australian Journal of Teacher Education, 34*(1), 42–53.

Wenger, E. (1998). *Communities of practice: Learning meaning and identity.* Cambridge University Press.

Wheeler, V. L. (1997). *Supplements to MNEMOSYNE: Stratagem and the vocabulary of military trickery.* Brill.

White Jr., L. (1964). *Medieval technology and social change.* Oxford University Press.

Wikipedia. (2020, July 1). *Assemblage*. Retrieved July 1, 2020, from https://en.wikipedia.org/wiki/Assemblage_(philosophy)

Williams, J. (1996). Philosophy of technology education. In J. Williams & A. Williams (Eds.), *Technology education for teachers* (pp. 27–62). Macmillan Education Australia.

Williams, J. (2010). Immanence. In A. Parr (Ed.), *The Deleuze dictionary: Revised edition* (pp. 128–130). Edinburgh University Press.

Williams, J. (2011). *Gilles Deleuze's philosophy of time* [eBook edition]. Edinburgh University Press.

Williams, J. (2013). *Gilles Deleuze's difference and repetition: A critical introduction and guide* (2nd ed.). Edinburgh University Press.

Williams, P. J. (2008). A global curriculum: Design technology in the international baccalaureate. *Design and Technology: An International Journal, 12*(3), 47–57.

Williams, P. J. (2009). Technological literacy: A multiliteracies approach for democracy. *International Journal of Technology and Design Education, 19*, 237–254.

Williams, P. J. (2018). Technology education: History of research. In M. J. de Vries (Ed.), *International handbook of technology education* (pp. 85–99). Springer.

Williams, P. J., Jones, A., & Buntting, C. (2015). *The future of technology education.* Springer.

Williams, R. J. (2014). *The Buddha in the machine: Art, technology, and the meeting of East and West.* Yale University Press.

Winner, L. (1992). *Autonomous technologies: Technics-out-of-control as a theme in political thought.* MIT Press. (Original work published 1977)

Wulf, C. (2013). *Anthropology: A continental perspective* (D. Winter, E. Hamilton, & M. Rouse, Trans.). University of Chicago Press.

# Index